Solar Energy Pocket Reference

ISES

International
Solar Energy
Society

Villa Tannheim
Wiesentalstr. 50
79115 Freiburg
Germany
www.ises.org

D0494028

earthscan

NOT TO BE
TAKEN AWAY

Preface

The ISES *Solar Energy Pocket Reference* contains a lot of useful information and data needed by solar energy professionals, researchers, designers, technicians and those generally interested in solar energy systems and applications. The pocket book contains data, equations, figures and plots for worldwide solar radiation, solar angles, sunpath and shading, material properties, solar thermal collectors, energy storage, photovoltaic systems design and applications, solar water and space heating, daylighting design data, solar dryer configurations, life cycle costing and units and conversion factors. This information is gathered from books and other published literature and reviewed by a number of experts including scientists, engineers, and architects. We wish to express our deep gratitude to all the reviewers of this pocket book.

This pocket reference book is the first in a series of pocket books to be published by ISES, such as Wind Energy and Biomass, etc. We hope this series will serve the practitioners of the various renewable energy fields.

D. Yogi Goswami, Ph.D., P.E.
President, ISES

ISBN 978-1-84407-306-1

9 781844 073061

ISBN: 978-1-84407-306-1

Earthscan publishes in association with the International Institute for Environment and Development.

Contents

Sun/earth geometric relationship 1
Solar angles 2-3
Solar time
    conversion, local ↔ solar 3
    sunrise/sunset 3
Declination angle, table 4
Equation of time, table 4
World standard time meridians 5
Sun path diagrams 6-13
Shading
    overhang and fin 14
    adjacent structure 15-16
Solar radiation
    worldwide map, horizontal surface data 17
    worldwide table, horizontal surface data 18-31
    tilted surface, estimation procedure 32-36
    ultraviolet component 36
Radiative properties of materials
    absorptivity & emissivity of common materials 37-39
    absorptivity & emissivity of selective surfaces 40
    absorptivity, angular variation 41
    reflectivity of reflector materials 41
    absorption, solar radiation in water 41
Transparent materials
    index of refraction 42
    collector cover material properties 43-44
Thermal collector overview 45
Non-concentrating thermal collectors
    performance parameters 46
    incidence angle modifier 46-47
    European standard parameters 48
Concentrating thermal collectors
    performance parameters 48
    incidence factors 49
Thermal energy storage
    formulations 50
    containment material, thermal conductivities 50
    sensible storage materials 51

latent storage materials	52
thermochemical storage materials	53
Photovoltaics	
maximum power condition	54
cell voltage/current output effects	55-56
system design	57-59
Water pumping, load data	60-61
Wire, electrical load rating	61
Water heating	
common configurations	62-64
load data	65
Space heating, load data	65
Solar heating system evaluation, f-chart method	66-70
Daylighting	
guidelines	70-71
sidelight sizing, lumen method	71-76
transmissivity, window glass	73
skylight sizing, lumen method	77-78
Solar drying, configurations	79
Life cycle costing	80-82
References	83-84
Conversion factors, constants	85-86
ISES, history	87-88

Sun/Earth Geometric Relationship

Annual motion of earth around the sun.

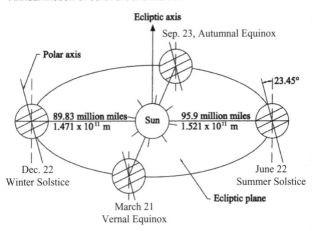

Location of the tropics.

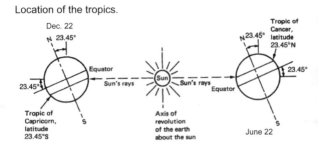

Solar Angle Formulations

Latitude angle, L, the angle between a line from the center of the earth to the site of interest and the equatorial plane. Values north of the equator are positive and those south are negative.

Solar declination, δ_s, the angle between the earth-sun line (through their centers) and the plane through the equator. Declinations are positive in the northern hemisphere and negative in the southern hemisphere.

$$\delta_s = 23.45° \sin\left(\frac{360(284 + n)}{365°}\right)$$

Where n is the day number (Jan. 1 = 1, Jan 2 = 2, etc.)

Hour angle, h_s, based on the nominal time of 24 hours for the sun to travel 360°, or 15° per hour. When the sun is due south (northern hemisphere, due north for southern hemisphere), the hour angle is 0, morning values are negative, and afternoon values are positive.

$$h_s = 15° \times (\text{hours from local solar noon})$$

$$= \frac{(\text{minutes from local solar noon})}{4 \, (\text{min}/\text{deg})}$$

Solar altitude angle, α, is the angle between a line collinear with the sun's rays and the horizontal plane.

$$\sin \alpha = \sin L \sin \delta_s + \cos L \cos h_s$$

Solar azimuth angle, a_s, is the angle between the projection of the earth-sun line on the horizontal plane and the due south direction (northern hemisphere) or due north (southern hemisphere).

$$a_s = \sin^{-1}\left(\frac{\cos \delta_s \sin h_s}{\cos \alpha}\right)$$

Note: $a_s > 90°$ that is $a_s \, (\text{am}/\text{pm}) = \mp\left(180° - |a_s|\right)$ when:

$$\begin{cases} L \leq \delta_s \\ L > \delta_s \text{ and solar time earlier than } t_E \text{ or later than } t_W \end{cases}$$

$$t_E / t_W = 12:00 \text{ noon} \mp \left(\cos^{-1} \left[\tan \delta_s / \tan L \right] \right) / 15 \, (\text{deg/hr})$$

Solar Time

Relation between local solar time and local standard time (LST):

$$\text{Solar Time} = \text{LST} + \text{ET} + \left(I_{st} - I_{local} \right) \times 4 \, \frac{\text{min}}{\text{deg}}$$

Where I_{st} is the standard time meridian, page 5, and I_{local} is the local longitude value. The equation of time, ET, can be gathered from tabular values, page 4, or computed with:

$$\text{ET} (\text{minutes}) = 9.87 \sin 2B - 7.53 \cos B - 1.5 \sin B$$

where:
$$B = \frac{360 (n - 81)}{364°}$$

At local solar noon:

$$h_s = 0 \qquad \alpha = 90° - \left| L - \delta_s \right| \qquad a_s = 0$$

Sunrise and Sunset Times

At sunrise or sunset the center of the sun is located at the eastern or western horizon, respectively. By definition, the altitude angle is then: $\alpha = 0$. The local solar time for sunrise/sunset can be computed with:

$$\text{Sunrise} / \text{Sunset} = 12:00 \text{ noon} \mp \left[\frac{\left(\cos^{-1} \left(-\tan L \times \tan \delta_s \right) \right)}{15 \, (\text{deg/hour})} \right]$$

For the tip of the sun at the horizon (apparent sunrise/sunset), subtract/add approximately 4 minutes from the sunrise/sunset times.

3

Values for Declination, δ_s, and Equation of Time, ET

Date	Declination Deg	Min	Equation of Time Min	Sec	Date	Declination Deg	Min	Equation of Time Min	Sec
Jan. 1	−23	4	−3	14	Feb. 1	−17	19	−13	34
5	22	42	5	6	5	16	10	14	2
9	22	13	6	50	9	14	55	14	17
13	21	37	8	27	13	13	37	14	20
17	20	54	9	54	17	12	15	14	10
21	20	5	11	10	21	10	50	13	50
25	19	9	12	14	25	9	23	13	19
29	18	9	123	5					
Mar. 1	−7	53	−12	38	Apr. 1	+4	14	−4	12
5	6	21	11	48	5	5	46	3	1
9	5	48	10	51	9	7	17	1	52
13	3	14	9	49	13	8	46	−0	47
17	1	39	8	42	17	10	12	+0	13
21	−0	5	7	32	21	11	35	1	6
25	+1	30	6	20	25	12	56	1	53
29	3	4	5	7	29	14	13	2	33
May 1	+14	50	+2	50	June 1	+21	57	2	27
5	16	2	34	17	5	22	28	1	49
9	17	9	3	35	9	22	52	1	6
13	18	11	3	44	13	23	10	+0	18
17	19	9	3	44	17	23	22	−0	33
21	20	2	3	24	21	23	27	1	25
25	20	49	3	16	25	23	25	2	17
29	21	30	2	51	29	23	17	3	7
July 1	+23	10	−3	31	Aug. 1	+18	14	−6	17
5	22	52	4	16	5	17	12	5	59
9	22	28	4	56	9	16	6	5	33
13	21	57	5	30	13	14	55	4	57
17	21	21	5	57	17	13	41	4	12
21	20	38	6	15	21	12	23	3	19
25	19	50	6	24	25	11	2	2	18
29	18	57	6	23	29	9	39	1	10
Sep. 1	+8	35	−0	15	Oct. 1	−2	53	+10	1
5	7	7	+1	2	5	4	26	11	17
9	5	37	2	22	9	5	58	12	27
13	4	6	3	45	13	7	29	13	30
17	2	34	5	10	17	8	58	14	25
21	1	1	6	35	21	10	25	15	10
25	0	32	8	0	25	11	50	15	46
29	2	6	9	22	29	13	12	16	10
Nov. 1	−14	11	+16	21	Dec. 1	−21	41	11	16
5	15	27	16	23	5	22	16	9	43
9	16	38	16	12	9	22	45	8	1
13	17	45	15	47	13	23	6	6	12
17	18	48	15	10	17	23	20	4	47
21	19	45	14	18	21	23	26	2	19
25	20	36	13	15	25	23	25	+0	20
29	21	21	11	59	29	23	17	−1	39

*Since each year is 365.25 days long, the precise value of declination varies from year to year. *The American Ephemeris and Nautical Almanac*, published each year by the U.S. Government Printing Office, contains precise values for each day of each year.

Worldwide standard time meridians

Relative to UTC	Description	Standard Meridian, l_{st} [°]
-11	Nome Time (US) Samoa Standard Time	165° W
-10	Hawaiian Standard Time (US) Tahiti Time	150° W
-9	Alaska Standard Time (US) Yukon Standard Time	135° W
-8	US Pacific Standard Time	120° W
-7	US Mountain Standard Time	105° W
-6	US Central Standard Time Mexico Time	90° W
-5	US Eastern Standard Time Colombia Time	75° W
-4	Atlantic Standard Time Bolivia Time	60° W
-3	Eastern Brazil Standard Time Argentina Time	45° W
-2	Greenland Eastern Std. Time Fernando de Noronha Time (Brazil)	30° W
-1	Azores Time Cape Verde Time	15° W
0	Greenwich Mean Time Coordinated Universal Time Western Europe Time	0°
+1	Central Europe Time Middle European Time	15° E
+2	Eastern Europe Time Kaliningrad Time (Russia)	30° E
+3	Moscow Time (Russia) Baghdad Time	45° E
+4	Volga Time (Russia) Gulf Standard Time	60° E
+5	Yekaterinburg Time (Russia) Pakistan Time	75° E
+5.5	Indian Standard Time	82.5° E
+6	Novosibirsk Time (Russia) Bangladesh Time	90° E
+7	Krasnoyarsk Time (Russia) Java Time	105° E
+8	Irkutsk Time (Russia) China Coast Time	120° E
+9	Yakutsk Time (Russia) Japan Standard Time	135° E
+10	Vladivostok Time (Russia) Guam Standard Time	150° E
+11	Magadan Time (Russia) Solomon Islands Time	165° E
±12	Kamchatka Time (Russia) New Zealand Standard Time	180°

Sun Path Diagrams

Procedure to determine the solar altitude and azimuth angles for a given latitude, time of year, and time of day:

1. **Transition the time of interest, local standard time (LST), to solar time.** See procedure page 3.

2. **Determine the declination angle based on time of year.** Use formulation page 2, or table page 4.

3. **Read the solar altitude and azimuth angles from the appropriate sun path diagram.** Diagrams are chosen based on latitude; linear interpolations are used for latitudes not covered. For values at southern latitudes change the sign of the solar declination.

Representative path diagram illustrating the determination of solar position.

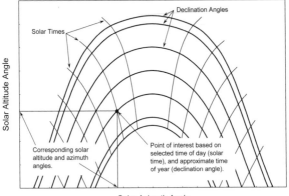

Solar Azimuth Angle

Note: The sign convention for the declination angle, δ_s, is for northern latitudes. To use the diagrams for southern latitudes, reverse the sign of the declination angle.

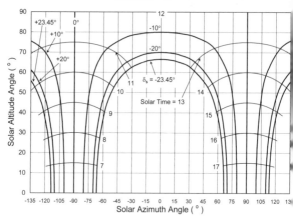

Latitude = 0°

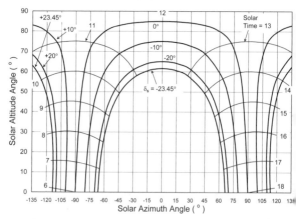

Latitude = 5°

Note: The sign convention for the declination angle, δ_s, is for northern latitudes. To use the diagrams for southern latitudes, reverse the sign of the declination angle.

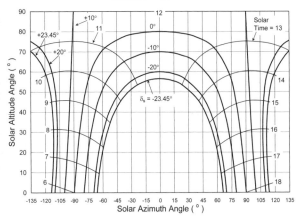

Latitude = 10°

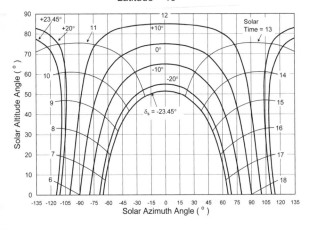

Latitude = 15°

Note: The sign convention for the declination angle, δ_s, is for northern latitudes. To use the diagrams for southern latitudes, reverse the sign of the declination angle.

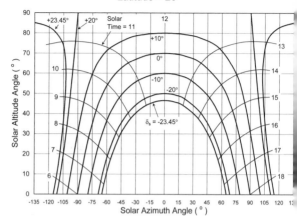

Latitude = 20°

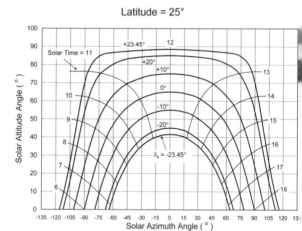

Latitude = 25°

Note: The sign convention for the declination angle, δ_s, is for northern latitudes. To use the diagrams for southern latitudes, reverse the sign of the declination angle.

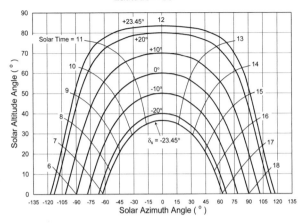

Latitude = 30°

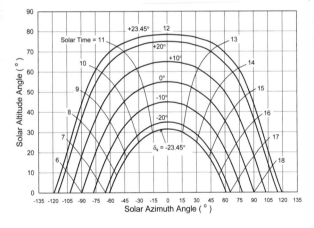

Latitude = 35°

Note: The sign convention for the declination angle, δ_s, is for northern latitudes. To use the diagrams for southern latitudes, reverse the sign of the declination angle.

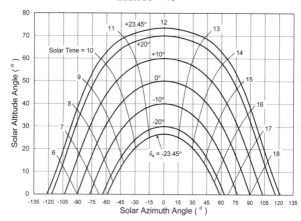

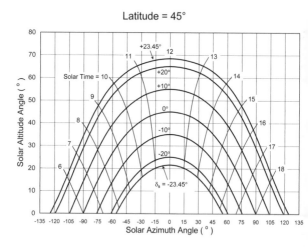

Note: The sign convention for the declination angle, δ_s, is for northern latitudes. To use the diagrams for southern latitudes, reverse the sign of the declination angle.

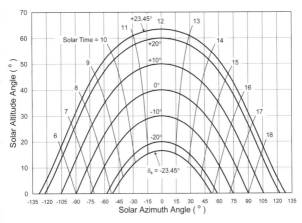

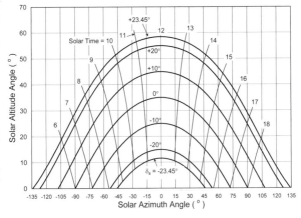

Note: The sign convention for the declination angle, δ_s, is for northern latitudes. To use the diagrams for southern latitudes, reverse the sign of the declination angle.

Latitude = 60°

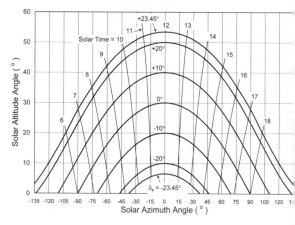

Latitude = 65°

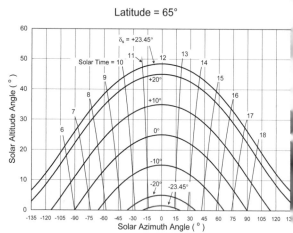

Note: The sign convention for the declination angle, δ_s, is for northern latitudes. To use the diagrams for southern latitudes, reverse the sign of the declination angle.

13

Overhang and Fin Shading Calculation

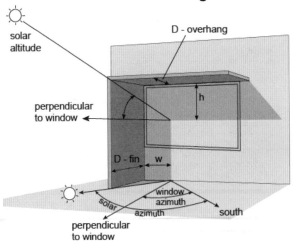

Reprinted from [3]

For overhang shading:

$$h = \frac{D \times \tan(\text{solar altitude})}{\cos(\text{solar azimuth} - \text{window azimuth})\,*}$$

For fin shading:

$$w = D \times \tan(\text{solar azimuth} - \text{window azimuth})\,*$$

Solar altitude and azimuth angles can be determined with the sun path diagrams, beginning on page 6.

*   Observe proper signs for both the solar and window azimuth values. Convention is that angles east of the equator vector are negative while those west are positive.

Obstruction Shading

Procedure for estimating shading caused by adjacent structures:

1. **Define the solid angle presented by the obstruction in terms of altitude, α_{obs}, and azimuth, a_{obs}, angles for the point of interest.** Choose characteristic points to describe the profile of the obstruction and determine their angular coordinates; refer to the geometry of the figures below and on the next page. By measuring or estimating the distances and heights in question, standard trigonometric formulas (next page) can be used to determine the angular values.

2. **Locate the solid angle with respect to the path of the sun.** Refer to the appropriate sun path diagram, based on the site latitude, and locate the solid angle on the diagram by plotting the α_{obs} and a_{obs} values on the solar altitude and solar azimuth axes, respectively. An example of this positioning is shown on the next page.

3. **Determine dates and times of shading.** On the sun path diagram, where the solid angle overlaps the declination lines, shading of the point of interest will occur. The dates can be estimated by reading the declination values that intersect the solid angle, and then matching the declination to the time of year, table page 4. In a similar manner, the hours of shading for a particular declination value, can be estimated by reading the solar time corresponding to the intersection of the declination of interest and the solid angle presented by the obstruction.

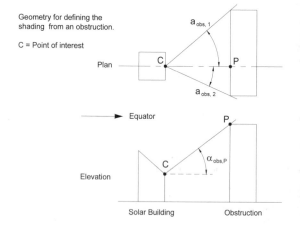

Geometry for defining the shading from an obstruction.

C = Point of interest

Obstruction Shading

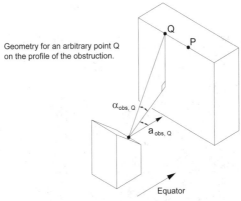

Geometry for an arbitrary point Q on the profile of the obstruction.

Example overlay of an obstruction profile on a sun path diagram.

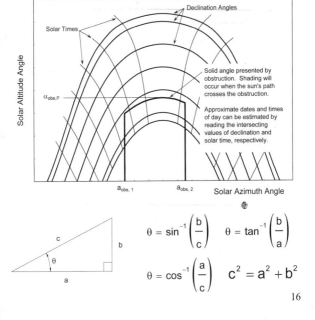

$$\theta = \sin^{-1}\left(\frac{b}{c}\right) \qquad \theta = \tan^{-1}\left(\frac{b}{a}\right)$$

$$\theta = \cos^{-1}\left(\frac{a}{c}\right) \qquad c^2 = a^2 + b^2$$

Annual average horizontal surface insolation

July 1983 – June 1993
Source: Surface meteorology and solar energy, National Aeronautics and Space Administration, USA; http://eosweb.larc.nasa.gov/sse

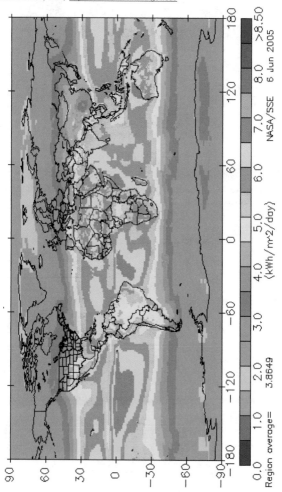

17

Worldwide, Horizontal Surface Solar Radiation Data, $[MJ/m^2\text{-}day]$

Position	Lat	Long	Jan.	Feb.	Mar.	Apr.	May	June	July	Aug.	Sep.	Oct.	Nov.	Dec.
Argentina														
Buenos Aires	34.58 S	58.48 W	24.86	21.75	18.56	11.75	8.71	7.15	7.82	8.75	14.49	16.66	24.90	21.93
Australia														
Adelaide	34.93 S	138.52 E	20.99	17.50	20.15	18.27	17.98	—	18.81	19.64	20.11	20.88	20.57	20.72
Brisbane	27.43 S	153.08 E	25.36	22.22	13.25	16.61	12.23	11.52	9.70	15.10	17.61	19.89	—	—
Canberra	35.30 S	148.18 E	28.20	24.68	20.56	14.89	10.29	6.62	9.41	12.33	16.88	24.06	26.00	25.77
Darwin	12.47 S	130.83 E	26.92	23.40	18.13	13.62	9.30	7.89	5.72	11.15	14.85	18.87	23.43	22.34
Hobart	42.88 S	147.32 E	—	—	—	10.09	7.26	6.04	6.54	9.21	13.54	18.12	—	—
Laverton	37.85 S	114.08 E	22.96	20.42	15.59	13.40	7.48	6.10	7.62	10.43	13.24	18.76	—	—
Sydney	33.87 S	151.20 E	21.09	21.75	17.63	13.63	9.78	8.79	—	12.84	16.93	22.10	—	—
Austria														
Wien	48.20 N	16.57 E	3.54	7.10	8.05	14.72	16.79	20.87	19.89	17.27	12.55	8.45	3.51	2.82
Innsbruck	47.27 N	11.38 E	5.57	9.28	10.15	15.96	14.57	17.65	18.35	17.26	12.98	9.08	4.28	3.50
Barbados														
Husbands	13.15 N	59.62 W	19.11	20.23	—	21.80	19.84	20.86	21.55	22.14	—	—	18.30	16.56
Belgium														
Ostende	51.23 N	2.92 E	2.82	5.75	9.93	15.18	16.74	16.93	18.21	18.29	11.71	6.15	2.69	1.97
Melle	50.98 N	3.83 E	2.40	4.66	8.41	13.55	14.23	13.28	15.71	15.61	10.63	5.82	2.40	1.59
Brunei														
Brunei	4.98 N	114.93 E	19.46	20.12	22.71	20.54	19.74	18.31	19.38	20.08	20.83	17.51	17.39	18.12
Bulgaria														
Chirpan	42.20 N	25.33 E	6.72	6.79	8.54	13.27	17.25	17.39	19.85	14.61	12.53	8.52	5.08	5.09
Sofia	42.65 N	23.38 E	4.05	6.23	7.93	9.36	12.98	19.73	19.40	17.70	14.71	6.44	—	3.14
Canada														
Montreal	45.47 N	73.75 E	4.74	8.33	11.84	10.55	15.05	22.44	21.08	18.67	14.83	9.18	4.04	4.01
Ottawa	45.32 N	75.67 E	5.34	9.59	13.33	13.98	20.18	20.34	19.46	17.88	13.84	7.38	4.64	5.04
Toronto	43.67 N	79.38 E	4.79	8.15	11.96	14.00	18.16	24.35	23.38	15.89	15.89	9.40	4.72	3.79
Vancouver	49.18 N	123.17 E	3.73	4.81	12.14	16.41	20.65	24.04	22.87	19.08	12.77	7.39	4.29	1.53
Chile														
Pascua	27.17 S	109.43 W	19.64	16.65	—	11.12	9.52	8.81	10.90	12.29	17.19	20.51	21.20	22.44
Santiago	33.45 S	70.70 W	18.61	16.33	13.44	8.32	5.07	3.66	3.35	5.65	8.15	13.62	20.14	23.88

Worldwide, Horizontal Surface Solar Radiation Data, [MJ/m²-day]

Position	Lat	Long	Jan.	Feb.	Mar.	Apr.	May	June	July	Aug.	Sep.	Oct.	Nov.	Dec.
China														
Beijing	39.93 N	116.28 W	7.73	10.59	13.87	17.93	20.18	18.65	15.64	16.61	15.52	11.29	7.25	6.89
Guangzhou	23.13 N	113.32 E	11.01	6.32	4.04	7.89	10.53	12.48	16.14	16.02	15.03	15.79	11.55	9.10
Harbin	45.75 N	126.77 E	5.15	9.54	17.55	20.51	20.33	17.85	19.18	16.09	13.38	14.50	10.50	6.98
Kunming	25.02 N	102.68 E	9.92	11.26	14.38	18.00	18.53	17.37	11.95	18.47	15.94	12.45	11.96	13.62
Lanzhou	36.05 N	103.88 E	7.30	12.47	10.62	18.91	17.40	20.40	20.23	17.37	13.23	10.21	8.22	6.43
Shanghai	31.17 N	121.43 E	7.44	10.31	11.78	14.36	14.23	16.79	14.63	11.85	15.96	12.03	7.73	8.70
Columbia														
Bogota	4.70 N	74.13 W	17.89	—	19.37	16.58	14.86	—	15.42	18.20	17.05	14.58	14.20	16.66
Cuba														
Havana	23.17 N	82.35 W	—	14.70	18.94	20.95	22.63	18.83	21.40	20.19	16.84	16.98	13.19	13.81
Czech														
Kucharovice	48.88 N	16.08 E	3.03	5.85	9.88	14.06	20.84	19.24	21.18	19.41	13.61	6.11	3.47	2.12
Churanov	49.07 N	13.62 E	2.89	5.82	9.24	13.18	21.32	15.68	20.51	19.49	12.84	5.68	3.36	2.99
Hradec Kralov	50.25 N	15.85 E	3.51	5.94	10.58	15.95	20.42	18.43	17.17	17.92	11.86	6.27	2.45	1.89
Denmark														
Copenhagen	55.67 N	12.30 E	1.83	3.32	7.09	11.12	21.39	24.93	—	13.92	10.10	5.20	2.81	1.23
Egypt														
Cairo	30.08 N	31.28 E	10.06	12.96	18.49	23.04	21.91	26.07	25.16	23.09	21.01	—	11.74	9.85
Mersa Matruh	31.33 N	27.22 E	8.38	11.92	18.47	24.27	24.17	—	26.67	26.27	21.92	18.28	11.71	8.76
Ethiopia														
Addis Ababa	8.98 N	38.80 E	—	11.39	—	12.01	—	—	—	6.33	9.35	11.71	11.69	11.50
Fiji														
Nandi	17.75 S	177.45 E	20.82	20.65	20.25	18.81	15.68	14.18	15.08	16.71	19.37	20.11	21.78	25.09
Suva	48.05 S	178.57 E	20.37	17.74	16.22	13.82	10.81	12.48	11.40	—	—	18.49	19.96	20.99
Finland														
Helsinki	60.32 N	24.97 E	1.13	2.94	5.59	11.52	17.60	16.81	20.66	15.44	8.44	3.31	0.97	0.63
France														
Agen	44.18 N	0.60 E	4.83	7.40	10.69	17.12	19.25	20.42	21.63	20.64	15.56	8.41	5.09	5.01
Nice	43.65 N	7.20 E	6.83	—	11.37	17.79	20.74	24.10	24.85	24.86	15.04	10.99	7.08	6.73
Paris	48.97 N	2.45 E	2.62	5.08	7.21	12.90	14.84	13.04	15.54	16.30	10.17	5.61	3.14	2.20

Worldwide, Horizontal Surface Solar Radiation Data, [MJ/m²-day]

Position	Lat	Long	Jan.	Feb.	Mar.	Apr.	May	June	July	Aug.	Sep.	Oct.	Nov.	Dec.
Germany														
Bonn	50.70 N	7.15 E	2.94	5.82	8.01	14.27	15.67	14.41	18.57	17.80	11.70	6.15	3.42	1.90
Nuremberg	53.33 N	13.20 E	3.23	6.92	9.08	15.69	15.71	18.21	21.14	17.98	12.43	8.15	2.79	2.51
Bremen	53.05 N	8.80 E	2.36	4.93	8.53	14.52	14.94	14.52	19.40	15.02	10.48	6.27	2.80	1.66
Hamburg	53.63 N	10.00 E	1.97	3.96	7.59	12.32	14.11	12.69	19.00	14.11	10.29	6.45	2.33	1.43
Stuttgart	48.83 N	9.20 E	3.59	7.18	9.22	15.81	17.72	17.44	22.21	19.87	12.36	7.81	3.19	2.54
Ghana														
Bole	9.03 N	2.48 W	18.29	19.76	19.71	19.15	16.61	—	—	13.68	16.29	17.27	17.33	15.93
Accra	5.60 N	0.17 W	14.82	16.26	18.27	16.73	18.15	13.96	13.86	13.49	15.32	19.14	18.16	14.23
Great Britain														
Belfast	54.65 N	6.22 W	2.00	3.60	6.85	12.00	15.41	15.09	15.46	13.56	11.49	4.63	2.34	1.24
Jersey	49.22 N	2.20 W	2.76	5.65	9.51	14.98	18.51	17.83	18.14	18.62	12.98	6.16	3.26	2.83
London	51.52 N	0.12 W	2.24	3.87	7.40	12.01	12.38	13.24	16.59	16.23	12.59	5.67	2.87	1.97
Greece														
Athens	37.97 N	23.72 E	9.11	10.94	15.70	20.91	23.85	25.48	24.21	23.08	19.03	13.29	5.98	6.64
Sikiwna	37.98 N	22.73 E	7.60	8.16	11.99	21.06	22.62	24.32	23.56	21.73	17.30	11.75	9.45	6.35
Guadeloupe														
Le Raizet	16.27 N	61.52 W	14.88	18.10	20.55	19.69	20.26	20.65	20.65	20.24	18.47	17.79	13.49	14.38
Guyana														
Cayenne	4.83 N	52.37 W	14.46	14.67	16.28	17.57	—	14.92	17.42	18.24	20.52	—	22.69	17.04
Hong Kong														
King's Park	22.32 N	114.17 W	12.34	7.39	6.94	9.50	11.38	13.60	16.70	17.06	15.91	16.52	14.19	10.00
Hungary														
Budapest	47.43 N	19.18 E	2.61	7.46	11.14	14.46	20.69	19.47	21.46	19.72	12.88	7.96	2.95	2.47
Iceland														
Reykjavik	64.13 N	21.90 W	0.52	2.02	6.25	11.77	13.07	14.58	16.83	11.35	9.70	3.18	1.00	0.65
India														
Bombay	19.12 N	72.85 E	18.44	21.00	22.72	24.52	24.86	19.75	15.84	16.00	18.19	20.38	19.18	17.81
Calcutta	22.53 N	88.33 E	15.69	18.34	20.09	22.34	22.37	17.55	17.07	16.55	16.52	16.90	16.35	15.00
Madras	13.00 N	80.18 E	19.09	22.71	25.14	24.88	23.89	—	18.22	19.68	19.81	16.41	14.76	15.79

Worldwide, Horizontal Surface Solar Radiation Data, [MJ/m²-day]

Position	Lat	Long	Jan.	Feb.	Mar.	Apr.	May	June	July	Aug.	Sep.	Oct.	Nov.	Dec.
India														
Nagpur	21.10 N	79.05 E	18.08	21.01	22.25	24.08	24.79	19.84	15.58	15.47	17.66	20.10	18.98	17.33
New Delhi	28.58 N	77.20 E	14.62	18.25	20.15	23.40	23.80	19.16	20.20	19.89	20.08	19.74	16.95	14.22
Ireland														
Dublin	53.43 N	6.25 W	2.51	4.75	7.48	11.06	17.46	19.11	15.64	13.89	9.65	5.77	2.93	—
Israel														
Jerusalem	31.78 N	35.22 E	10.79	13.01	18.08	23.79	29.10	31.54	31.83	28.79	25.19	20.26	12.61	10.71
Italy														
Milan	45.43 N	9.28 E	—	6.48	10.09	13.17	17.55	16.32	18.60	16.86	11.64	5.40	3.52	2.41
Rome	41.80 N	12.55 E	—	9.75	13.38	15.82	15.82	18.89	22.27	21.53	16.08	8.27	6.41	4.49
Japan														
Fukuoka	33.58 N	130.38 E	8.11	8.72	10.95	13.97	14.36	12.81	13.84	16.75	13.92	11.86	10.05	7.30
Tateno	36.05 N	140.13 E	9.06	12.17	11.00	15.78	16.52	15.26	—	14.12	—	9.60	8.55	8.26
Yonago	35.43 N	133.35 E	6.25	7.16	10.87	17.30	16.72	15.44	17.06	19.93	12.41	10.82	7.50	5.51
Kenya														
Mombasa	4.03 S	39.62 E	22.30	22.17	22.74	18.49	18.31	17.41	—	18.12	21.03	22.97	21.87	21.25
Nairobi	1.32 S	36.92 E	—	24.10	21.20	18.65	14.83	15.00	13.44	14.12	19.14	19.38	16.90	18.27
Lithuania														
Kaunas	54.88 N	23.88 E	1.89	4.43	7.40	12.97	18.88	18.74	21.41	15.79	10.40	5.64	1.80	1.10
Madagascar														
Antananarivo	18.80 S	47.48 E	15.94	13.18	13.07	11.53	9.25	8.21	9.32	—	—	16.43	15.19	15.62
Malaysia														
Kualalumpur	3.12 N	101.55 E	15.36	17.67	18.48	16.87	15.67	16.24	15.32	15.89	14.62	14.13	13.54	11.53
Piang	5.30 N	100.27 E	19.47	21.35	23.24	20.52	18.63	19.32	17.17	16.96	15.93	16.01	18.35	17.37
Martinique														
Le Lamentin	14.60 N	61.00 W	17.76	20.07	22.53	21.95	22.42	21.23	20.86	21.84	20.23	19.87	14.08	16.25
Mexico														
Chihuahua	28.63 N	106.08 W	14.80	—	—	—	26.94	26.28	24.01	24.22	20.25	19.55	10.57	15.79
Orizabita	20.58 N	99.20 E	19.49	23.07	27.44	27.35	26.04	25.05	—	27.53	21.06	17.85	15.48	12.93

Worldwide, Horizontal Surface Solar Radiation Data, [MJ/m²-day]

Position	Lat	Long	Jan.	Feb.	Mar.	Apr.	May	June	July	Aug.	Sep.	Oct.	Nov.	Dec.
Mongolia														
Ulan Bator	47.93 N	106.98 E	6.28	9.22	14.34	18.18	20.50	19.34	16.34	16.65	14.08	11.36	7.19	5.35
Uliassutai	47.75 N	96.85 E	6.43	10.71	14.83	20.32	23.86	20.46	21.66	17.81	15.97	10.92	7.32	5.08
Morocco														
Casablanca	33.57 N	7.67 E	11.46	12.70	15.93	21.25	24.45	25.27	25.53	23.60	19.97	14.68	11.61	9.03
Mozambique														
Maputo	25.97 S	32.60 E	26.35	23.16	19.33	20.54	16.33	14.17	—	—	—	22.55	25.48	26.19
Netherlands														
Maastricht	50.92 N	5.78 E	3.20	5.43	8.48	14.82	14.97	14.32	18.40	17.51	11.65	6.51	3.01	1.72
New Caledonia														
Koumac	20.57 S	164.28 E	24.89	21.15	16.96	18.98	15.67	14.55	15.75	17.62	22.48	15.83	27.53	26.91
New Zealand														
Wilmington	41.28 S	174.77 E	22.59	19.67	14.91	9.52	6.97	4.37	5.74	7.14	12.50	16.34	19.07	24.07
Christchurch	43.48 S	172.55 E	23.46	19.68	13.98	8.96	6.47	4.74	5.38	6.94	13.18	17.45	18.91	24.35
Nigeria														
Benin City	6.32 N	5.60 E	14.89	17.29	19.15	17.21	16.97	15.04	10.24	12.54	14.37	15.99	17.43	15.75
Norway														
Bergen	60.40 N	5.32 E	0.46	1.33	3.18	8.36	19.24	16.70	16.28	10.19	6.53	3.19	1.36	0.35
Oman														
Seeb	23.58 N	58.28 E	12.90	14.86	21.22	22.22	25.30	24.02	23.46	21.66	20.07	18.45	15.49	13.12
Salalah	17.03 N	54.08 E	16.52	16.92	18.49	20.65	21.46	16.92	8.52	11.41	17.14	18.62	16.42	—
Pakistan														
Karachi	24.90 N	67.13 E	13.84	—	—	19.69	20.31	16.62	—	—	—	—	12.94	11.07
Multan	30.20 N	71.43 E	12.29	15.86	18.33	22.35	22.57	21.65	20.31	20.44	20.57	15.91	12.68	10.00
Islamabad	33.62 N	73.10 E	10.38	12.42	16.98	22.65	—	25.49	20.64	18.91	14.20	15.30	10.64	8.30
Peru														
Puno	15.83 S	70.02 W	14.98	12.92	16.08	20.03	17.45	17.42	15.74	15.32	16.11	16.18	14.24	13.90
Poland														
Warszawa	52.28 N	20.97 E	1.73	3.83	7.81	10.53	19.22	17.11	20.18	15.00	10.65	4.95	2.39	1.68
Kolobrzeg	54.18 N	15.58 E	2.50	3.25	8.86	15.21	20.79	20.50	17.19	16.46	7.95	5.75	1.78	1.18

Worldwide, Horizontal Surface Solar Radiation Data, [MJ/m²-day]

Position	Lat	Long	Jan.	Feb.	Mar.	Apr.	May	June	July	Aug.	Sep.	Oct.	Nov.	Dec.
Portugal														
Evora	38.57 N	7.90 W	9.92	12.43	17.81	18.69	23.57	29.23	28.75	23.77	20.17	—	6.81	4.57
Lisbon	38.72 N	9.15 W	9.24	11.60	17.52	18.49	24.64	29.02	28.14	22.20	19.76	13.56	7.18	4.83
Romania														
Bucuresti	44.50 N	26.13 E	7.05	10.22	12.04	16.53	18.97	22.16	23.19	—	17.17	9.55	4.82	—
Constanta	44.22 N	28.63 E	5.62	9.28	14.31	20.59	23.23	25.80	27.98	24.22	16.91	11.89	6.19	5.10
Galati	45.50 N	28.02 E	6.09	9.33	14.31	17.75	21.77	22.74	25.55	19.70	14.05	11.26	6.32	5.38
Russia														
Alexandovsko	60.38 N	77.87 E	1.34	4.17	9.16	17.05	21.83	21.34	20.26	13.05	10.16	4.68	1.71	0.68
Moscow	55.75 N	37.57 E	1.45	3.96	8.09	11.69	18.86	18.12	17.51	14.17	10.92	4.03	2.28	1.29
St. Petersburg	59.97 N	30.30 E	1.03	3.11	4.88	12.24	20.59	21.55	20.43	13.27	7.83	2.93	1.16	0.59
Verkhoyansk	67.55 N	133.38 E	0.21	2.25	7.61	15.96	19.64	—	—	14.12	7.59	3.51	0.54	—
St. Pierre & Miquelon														
St. Pierre	46.77 N	56.17 W	4.43	6.61	12.50	17.57	18.55	17.84	19.95	16.46	12.76	8.15	3.69	3.33
Singapore														
Singapore	1.37 N	103.98 E	19.08	20.94	20.75	18.20	14.89	15.22	13.92	16.66	16.51	15.82	13.81	12.67
South Korea														
Seoul	37.57 N	126.97 E	6.24	9.40	10.34	13.98	16.35	17.49	10.65	12.94	11.87	10.35	6.47	5.14
South Africa														
Cape Town	33.98 S	18.60 E	27.47	25.57	—	15.81	11.44	9.08	8.35	13.76	17.30	22.16	26.37	27.68
Port Elizabeth	33.98 S	25.60 E	27.22	22.06	19.01	15.29	11.79	11.13	10.73	13.97	18.52	23.09	23.15	27.26
Pretoria	25.73 S	28.18 E	26.06	22.43	20.52	16.09	15.67	13.67	15.19	18.65	21.62	21.75	24.82	23.43
Spain														
Madrid	40.45 N	3.72 W	7.73	10.53	15.35	21.74	22.81	22.05	26.27	22.90	18.89	10.21	8.69	5.56
Sudan														
Wad Madani	14.40 N	33.48 E	21.92	24.01	23.43	25.17	23.92	23.51	22.40	22.85	21.75	20.47	20.19	19.21
Elfasher	13.62 N	25.33 E	21.56	21.84	24.54	25.29	24.31	24.15	22.87	21.19	22.58	23.85	—	—
Shambat	15.67 N	32.53 E	23.90	27.38	—	27.45	23.21	26.15	23.55	25.46	24.05	23.51	23.82	22.53
Sweden														
Karlstad	59.37 N	13.47 E	1.26	3.13	5.02	14.01	19.90	16.70	20.92	14.14	10.52	3.98	1.47	0.94

Worldwide, Horizontal Surface Solar Radiation Data, [MJ/m²-day]

Position	Lat	Long	Jan.	Feb.	Mar.	Apr.	May	June	July	Aug.	Sep.	Oct.	Nov.	Dec.
Sweden														
Lund	55.72 N	13.22 E	1.97	3.47	6.66	12.48	17.83	13.38	18.74	14.99	10.39	5.45	1.82	1.21
Stockholm	59.35 N	18.07 E	1.32	2.69	4.75	13.21	15.58	14.79	20.52	14.48	10.50	4.04	1.19	0.83
Switzerland														
Geneva	46.25 N	6.13 E	2.56	7.21	9.46	17.07	20.98	19.78	22.38	20.50	13.62	8.44	3.31	2.87
Zurich	47.48 N	8.53 E	2.31	7.02	7.54	15.04	16.33	16.73	20.28	18.32	12.52	7.18	2.64	2.29
Thailand														
Bangkok	13.73 N	100.57 E	16.67	19.34	23.00	22.48	20.59	17.71	18.02	16.04	16.23	16.81	18.60	16.43
Trinidad & Tobago														
Crown Point	11.15 N	60.83 W	13.05	15.61	15.17	16.96	17.61	15.37	13.16	13.08	12.24	8.76	—	—
Tunisia														
Sidi Bouzid	36.87 N	10.35 E	7.88	10.38	13.20	17.98	25.12	26.68	27.43	24.33	18.87	12.11	9.37	6.72
Tunis	36.83 N	10.23 E	7.64	9.88	14.79	31.61	25.31	26.03	26.60	20.37	19.58	12.91	9.35	7.16
Ukraine														
Kiev	50.40 N	30.45 E	2.17	4.87	11.15	12.30	20.49	—	18.99	18.55	9.72	9.84	3.72	2.52
Uzbekistan														
Tashkent	41.27 N	69.27 E	7.27	10.81	15.93	23.60	25.21	29.53	28.50	26.68	20.76	13.25	8.61	4.59
Venezuela														
Caracas	10.50 N	66.88 W	14.25	13.56	16.30	15.56	15.69	15.56	16.28	17.11	17.04	15.14	14.74	13.50
St. Antonio	7.85 N	72.45 W	11.78	10.54	10.65	12.07	12.65	21.20	14.68	15.86	16.62	15.32	12.28	11.28
St. Fernando	7.90 N	67.42 W	14.92	16.82	16.89	—	—	14.09	13.78	14.42	14.86	15.27	14.25	13.11
Vietnam														
Hanoi	21.03 N	105.85 E	5.99	7.48	8.73	13.58	19.10	21.26	19.85	19.78	20.67	14.78	12.44	13.21

Worldwide, Horizontal Surface Solar Radiation Data, [MJ/m²-day]

Position	Lat	Long	Jan.	Feb.	Mar.	Apr.	May	June	July	Aug.	Sep.	Oct.	Nov.	Dec.
Yugoslavia														
Beograd	44.78 N	20.53 E	4.92	6.27	10.64	14.74	20.95	22.80	22.09	20.27	15.57	11.24	6.77	4.99
Kopaonik	43.28 N	20.80 E	7.03	10.93	14.75	12.78	13.54	20.43	22.48	—	20.14	11.61	6.26	4.64
Portoroz	45.52 N	13.57 E	5.11	7.84	13.75	17.30	23.66	22.31	25.14	21.34	13.40	8.98	6.04	3.92
Zambia														
Lusaka	15.42 S	28.32 W	16.10	18.02	20.24	19.84	17.11	16.37	19.45	20.72	21.68	23.83	23.85	20.52
Zimbabwe														
Bulawayo	20.15 S	28.62 N	20.03	22.11	21.03	18.09	17.15	15.36	16.46	19.49	21.55	23.44	25.08	23.46
Harare	17.83 S	31.02 N	19.38	19.00	19.22	17.67	18.35	16.10	14.55	17.87	21.47	23.98	19.92	21.88

(Source: Voeikov Main Geophysical Observatory, Russia: Internet address: http://wrdc-mgo.nrel.gov/html/get_data-ap.html)

Note: Data for 872 locations is available from these sources in 68 countries.

*Source for Canadian Data: Environment Canada: Internet address: http://www.ec-gc.ca./envhome.html.

25

Worldwide, Horizontal Surface Solar Radiation Data, [MJ/m²-day]

Position	Jan.	Feb.	Mar.	Apr.	May	June	July	Aug.	Sep.	Oct.	Nov.	Dec.	Average
Alabama													
Birmingham	9.20	11.92	15.67	19.65	21.58	22.37	21.24	20.21	17.15	14.42	10.22	8.40	16.01
Montgomery	9.54	12.49	16.24	20.33	22.37	23.17	21.80	20.56	17.72	14.99	10.90	8.97	16.58
Alaska													
Fairbanks	0.62	2.77	8.31	14.66	17.98	19.65	16.92	12.36	7.02	3.20	1.01	0.23	8.74
Anchorage	1.02	3.41	8.18	13.06	15.90	17.72	16.69	12.72	8.06	3.97	1.48	0.56	8.63
Nome	0.51	2.95	8.29	15.22	18.97	19.65	16.69	11.81	7.72	3.63	0.99	0.09	8.86
St. Paul Island	1.82	4.32	8.52	12.72	14.08	14.42	12.83	10.33	7.84	4.54	2.16	1.25	7.95
Yakutat	1.36	3.63	7.72	12.61	14.76	15.79	14.99	12.15	7.95	3.97	1.82	0.86	8.18
Arizona													
Phoenix	11.58	15.33	19.87	25.44	28.85	30.09	27.37	25.44	21.92	17.60	12.95	10.56	20.56
Tucson	12.38	15.90	20.21	25.44	28.39	29.30	25.44	24.08	21.58	17.94	13.63	11.24	20.44
Arkansas													
Little Rock	9.09	11.81	15.56	19.19	21.80	23.51	23.17	21.35	17.26	14.08	9.77	8.06	16.24
Fort Smith	9.31	12.15	15.67	19.31	21.69	23.39	23.85	24.46	17.26	13.97	9.88	8.29	16.35
California													
Bakersfield	8.29	11.92	16.69	22.15	26.57	28.96	28.73	26.01	21.35	15.90	10.33	7.61	18.74
Fresno	7.61	11.58	16.81	22.49	27.14	29.07	28.96	25.89	21.12	15.56	9.65	6.70	18.62
Long Beach	9.99	12.95	17.03	21.60	23.17	24.19	26.12	24.08	19.31	14.99	11.24	9.31	17.83
Sacramento	6.93	10.68	15.56	21.24	25.89	28.28	28.62	25.32	20.56	14.54	8.63	6.25	17.72
San Diego	11.02	13.97	17.72	21.92	22.49	23.28	24.98	23.51	19.53	15.79	12.26	10.22	18.06
San Francisco	7.72	10.68	15.22	20.44	24.08	25.78	26.46	23.39	19.31	13.97	8.97	7.04	16.92
Los Angeles	10.11	13.06	17.26	21.80	23.05	23.74	25.67	23.51	18.97	14.99	11.36	9.31	17.72
Santa Maria	10.22	13.29	17.49	22.26	25.10	26.57	26.91	24.42	20.10	15.67	11.47	9.54	18.62
Colorado													
Boulder	7.84	10.45	15.64	17.94	17.94	20.47	20.28	17.12	16.07	12.09	8.66	7.10	14.31
Colorado Springs	9.09	12.15	16.13	20.33	22.26	24.98	23.96	21.69	18.51	14.42	9.99	8.18	16.81
Connecticut													
Hartford	6.70	9.65	13.17	16.69	19.53	21.24	21.12	18.51	14.76	10.68	6.59	5.45	13.74

Worldwide, Horizontal Surface Solar Radiation Data, [MJ/m²-day]

Position	Jan.	Feb.	Mar.	Apr.	May	June	July	Aug.	Sep.	Oct.	Nov.	Dec.	Average
Delaware													
Wilmington	7.27	10.22	13.97	17.60	20.33	22.49	21.80	19.65	15.79	11.81	7.84	6.25	14.65
Florida													
Daytona Beach	11.24	13.85	17.94	22.15	23.17	22.03	21.69	20.44	17.72	14.99	12.15	10.33	17.38
Jacksonville	10.45	13.17	17.03	21.12	22.03	21.58	21.01	19.42	16.69	14.20	11.47	9.65	16.47
Tallahassee	10.33	13.29	16.92	21.24	22.49	22.03	20.90	19.65	17.72	15.56	11.92	9.77	16.81
Miami	12.72	15.22	18.51	21.58	21.46	20.10	21.10	20.10	17.60	15.67	13.17	11.81	17.38
Key West	13.17	16.01	19.65	22.71	22.83	22.03	22.03	21.01	18.74	16.47	13.85	15.79	18.40
Tampa	11.58	14.42	18.17	22.26	23.05	21.92	20.90	19.65	17.60	16.01	12.83	11.02	17.49
Georgia													
Athens	9.43	12.38	16.01	20.21	22.03	22.83	21.80	20.21	17.26	14.42	10.45	8.40	16.29
Atlanta	9.31	12.26	16.13	20.33	22.37	23.17	22.15	20.56	17.49	14.54	10.56	8.52	16.43
Columbus	9.77	12.72	16.47	20.67	22.37	22.83	21.58	20.33	17.60	14.99	11.02	9.09	16.62
Macon	9.54	12.61	16.35	20.56	22.37	22.83	21.58	20.21	17.26	14.88	10.90	8.86	16.50
Savanna	9.99	12.72	16.81	21.01	22.37	22.60	21.80	19.76	16.92	14.65	11.13	9.20	16.58
Hawaii													
Honolulu	14.08	16.92	19.42	21.24	22.83	23.51	23.74	23.28	21.35	18.06	14.88	13.40	19.42
Idaho													
Boise	5.79	8.97	13.63	18.97	23.51	26.01	27.37	23.62	18.40	12.26	6.70	5.11	15.90
Illinois													
Chicago	6.47	9.31	12.49	16.47	20.44	22.60	22.03	19.31	15.10	10.79	6.47	5.22	13.85
Rockford	6.70	9.77	12.72	16.58	20.33	22.49	22.15	19.42	15.22	10.79	6.59	5.34	14.08
Springfield	7.50	10.33	13.40	17.83	21.46	23.51	23.05	20.56	16.58	12.26	7.72	6.13	15.10
Indiana													
Indianapolis	7.04	9.99	13.17	17.49	21.24	23.28	22.60	20.33	16.35	11.92	7.38	5.79	14.76
Iowa													
Mason City	6.70	9.77	13.29	16.92	20.78	22.83	22.71	19.76	15.33	10.90	6.59	5.45	14.31
Waterloo	6.81	9.77	13.06	16.92	20.56	22.83	22.60	19.76	15.33	10.90	6.70	5.45	14.20

Worldwide, Horizontal Surface Solar Radiation Data, [MJ/m²-day]

Position	Jan.	Feb.	Mar.	Apr.	May	June	July	Aug.	Sep.	Oct.	Nov.	Dec.	Average
Kansas													
Dodge City	9.65	12.83	16.69	21.01	23.28	25.78	25.67	22.60	18.40	14.42	10.11	8.40	17.49
Goodland	8.97	11.92	16.13	20.44	22.71	25.78	25.55	22.60	18.28	14.08	9.65	7.84	17.03
Kentucky													
Lexington	7.27	9.88	13.51	17.60	20.56	22.26	21.46	19.65	16.01	12.38	7.95	6.25	14.54
Louisville	7.27	10.22	13.63	17.83	20.90	22.71	22.03	20.10	16.35	12.38	7.95	6.25	14.76
Louisiana													
New Orleans	9.77	12.83	16.01	19.87	21.80	22.03	20.67	19.65	17.60	15.56	11.24	9.31	16.35
Lake Charles	9.77	12.83	16.13	19.31	21.58	22.71	21.58	20.33	18.06	15.56	11.47	9.31	16.58
Maine													
Portland	6.70	9.99	13.78	16.92	19.99	21.92	21.69	19.31	15.22	10.56	6.47	5.45	13.97
Maryland													
Baltimore	7.38	10.33	13.97	17.60	20.21	22.15	21.69	19.19	15.79	11.92	8.06	6.36	14.54
Massachusetts													
Boston	6.70	9.65	13.40	16.92	20.21	22.03	21.80	19.31	15.33	10.79	6.81	5.45	14.08
Michigan													
Detroit	5.91	8.86	12.38	16.47	20.33	22.37	21.92	18.97	14.76	10.11	6.13	4.66	13.63
Lansing	5.91	8.86	12.49	16.58	20.21	22.26	21.92	18.85	14.54	9.77	5.91	4.66	13.51
Minnesota													
Duluth	5.68	9.31	13.74	17.38	20.10	21.46	21.80	18.28	13.29	8.86	5.34	4.43	13.29
Minneapolis	6.36	9.77	13.51	16.92	20.56	22.49	22.83	19.42	14.65	9.99	6.13	4.88	13.97
Rochester	6.36	9.65	13.17	16.58	20.10	22.15	22.15	19.08	14.54	10.11	6.25	5.11	13.74
Mississippi													
Jackson	9.43	12.38	16.13	19.87	22.15	23.05	22.15	19.08	14.54	10.11	6.25	5.11	13.74
Missouri													
Columbia	8.06	10.90	14.31	18.62	21.58	23.62	23.85	21.12	16.69	12.72	8.29	6.70	15.56
Kansas City	7.95	10.68	14.08	18.28	21.24	23.28	23.62	20.78	16.58	12.72	8.40	6.70	15.44
Springfield	8.52	11.02	14.65	18.62	21.24	23.05	23.62	21.24	16.81	13.17	8.86	7.27	15.67
St. Louis	7.84	10.56	13.97	18.06	21.12	23.05	22.94	20.44	16.58	12.49	8.18	6.59	15.22

Worldwide, Horizontal Surface Solar Radiation Data, [MJ/m²-day]

Position	Jan.	Feb.	Mar.	Apr.	May	June	July	Aug.	Sep.	Oct.	Nov.	Dec.	Average
Montana													
Helena	5.22	8.29	12.61	17.15	20.67	23.28	25.21	21.24	15.79	10.45	6.02	4.43	14.20
Lewistown	5.22	8.40	12.72	17.15	20.33	23.05	24.53	20.78	15.10	10.22	5.91	4.32	13.97
Nebraska													
Omaha	7.50	10.33	13.97	18.06	21.24	2.40	23.51	20.56	16.01	11.81	7.61	6.13	15.10
Lincoln	7.33	10.10	13.65	16.22	19.26	21.21	22.15	18.87	15.44	11.54	7.76	6.20	14.16
Nevada													
Elko	7.61	10.56	14.42	18.85	22.71	25.67	26.69	23.62	19.31	13.63	8.29	6.70	16.58
Las Vegas	10.79	14.42	19.42	24.87	28.16	30.09	28.28	25.89	22.15	17.03	12.15	9.88	20.33
Reno	8.29	11.58	16.24	21.24	25.10	27.48	28.16	24.98	20.56	14.88	9.31	7.38	17.94
New Hampshire													
Concord	6.81	10.11	13.97	16.92	20.21	21.80	21.80	19.08	14.99	10.45	6.47	5.45	14.08
New Jersey													
Atlantic City	7.38	10.22	13.97	17.49	20.21	21.92	21.24	19.19	15.79	11.92	8.06	6.36	14.54
Newark	6.93	9.77	13.51	17.26	19.76	21.35	21.01	18.85	15.33	11.36	7.27	5.68	13.97
New Mexico													
Albuquerque	11.47	14.99	19.31	24.53	27.60	29.07	27.03	24.76	21.12	17.03	12.49	10.33	19.99
New York													
Albany	6.36	9.43	12.95	16.69	19.53	21.46	21.58	18.51	14.65	10.11	6.13	5.00	13.51
Buffalo	5.68	8.40	12.15	16.35	19.76	22.03	21.69	18.62	14.08	9.54	5.68	4.54	13.29
New York City	6.93	9.88	13.85	17.72	20.44	22.03	21.69	19.42	15.56	11.47	7.27	5.79	14.31
Rochester	5.68	8.52	12.26	16.58	19.87	21.92	21.69	18.51	14.20	9.54	5.68	4.54	13.29
North Carolina													
Charlotte	8.97	11.81	15.67	19.76	21.58	22.60	21.92	19.99	16.92	13.97	9.99	8.06	16.01
Wilmington	9.31	12.15	16.24	20.44	21.92	22.60	21.58	19.53	16.69	14.08	10.56	8.52	16.13
North Dakota													
Fargo	5.79	9.09	13.17	16.92	20.56	22.37	23.17	19.87	14.31	9.54	5.68	4.54	13.74
Bismarck	6.12	9.75	13.88	17.43	21.45	23.01	24.06	20.12	15.21	10.61	6.28	4.84	14.39

Worldwide, Horizontal Surface Solar Radiation Data, [MJ/m²-day]

Position	Jan.	Feb.	Mar.	Apr.	May	June	July	Aug.	Sep.	Oct.	Nov.	Dec.	Average
Ohio													
Cleveland	5.79	8.63	12.04	16.58	20.10	22.15	21.92	18.97	14.76	10.22	6.02	4.66	13.51
Columbus	6.47	9.09	12.49	16.58	19.76	21.58	21.12	18.97	15.44	11.24	6.81	5.34	13.74
Dayton	6.81	9.43	12.83	17.03	20.33	22.37	22.37	19.65	15.90	11.47	7.04	5.45	14.20
Youngstown	5.79	8.40	11.92	15.90	19.19	21.24	20.78	18.06	14.31	10.11	6.02	4.77	13.06
Oklahoma													
Oklahoma City	9.88	1.25	16.47	20.33	22.26	24.42	24.98	22.49	18.17	14.54	10.45	8.74	17.15
Oregon													
Eugene	4.54	7.04	11.24	15.79	19.99	22.37	24.19	21.01	15.90	9.65	5.11	3.75	13.40
Medford	5.34	8.52	13.17	18.62	23.39	26.23	27.82	23.96	18.62	11.92	6.02	4.43	15.67
Portland	4.20	6.70	10.68	15.10	18.97	21.24	22.60	19.53	14.88	9.20	4.88	3.52	12.61
Pacific Islands													
Guam	16.35	17.38	19.65	20.78	20.56	19.76	18.28	17.49	17.49	16.58	15.79	15.10	17.94
Pennsylvania													
Philadelphia	7.04	9.88	13.63	17.26	19.99	22.03	21.46	19.42	15.67	11.58	7.72	6.02	14.31
Pittsburgh	6.25	8.97	12.61	16.47	19.65	21.80	21.35	18.85	15.10	10.90	6.59	5.00	13.63
Rhode Island													
Providence	6.70	9.65	13.40	16.92	19.99	21.58	21.24	18.85	15.22	11.02	6.93	5.56	13.97
South Carolina													
Charleston	9.77	12.72	16.81	21.12	22.37	22.37	21.92	19.65	16.92	14.54	11.02	9.09	16.58
Greenville	9.20	12.04	15.90	19.99	21.58	22.60	21.58	19.87	16.81	14.08	10.22	8.18	16.01
South Dakota													
Pierre	6.47	9.54	13.85	17.94	21.46	24.08	24.42	21.46	16.35	11.24	7.04	5.45	14.99
Rapid City	6.70	9.88	14.20	18.28	21.46	24.19	24.42	21.80	16.92	11.81	7.50	5.79	15.33
Tennessee													
Memphis	8.86	11.58	15.22	19.42	22.03	23.85	23.39	21.46	17.38	14.20	9.65	7.84	16.24
Nashville	8.29	11.13	14.65	19.31	21.69	23.51	22.49	20.56	16.81	13.51	8.97	7.15	15.67

Worldwide, Horizontal Surface Solar Radiation Data, [MJ/m²-day]

Position	Jan.	Feb.	Mar.	Apr.	May	June	July	Aug.	Sep.	Oct.	Nov.	Dec.	Average
Texas													
Austin	10.68	13.63	17.03	19.53	21.24	23.74	24.42	22.83	18.85	15.67	11.92	9.99	17.49
Brownsville	10.33	13.17	16.47	19.08	20.78	22.83	23.28	21.58	18.62	16.13	12.38	9.88	17.03
El Paso	12.38	16.24	20.90	25.44	28.05	28.85	26.46	24.30	21.12	17.72	13.63	11.47	20.56
Houston	9.54	12.26	15.22	18.06	20.21	21.69	21.35	20.21	17.49	15.10	11.02	8.97	15.90
San Antonio	10.88	13.53	16.26	17.35	21.10	23.87	24.92	22.81	19.22	15.52	11.50	9.98	17.24
Utah													
Salt Lake City	6.93	10.45	14.76	19.42	23.39	26.46	26.35	23.39	18.85	13.29	8.06	6.02	16.47
Vermont													
Burlington	5.79	9.20	13.06	16.47	19.87	21.69	21.80	18.74	14.42	9.43	5.56	4.43	13.40
Virginia													
Norfolk	8.06	10.90	14.65	18.51	20.78	22.15	21.12	19.42	16.13	12.49	9.09	7.27	15.10
Richmond	8.06	10.90	14.76	18.62	20.90	22.49	21.58	19.53	16.24	12.61	8.97	7.15	15.22
Washington													
Olympia	3.63	6.02	9.99	14.20	18.06	20.10	21.12	18.17	13.63	7.95	4.32	3.07	11.70
Seattle	3.52	5.91	10.11	14.65	19.08	20.78	21.80	18.51	13.51	7.95	4.20	2.84	11.92
Yakima	4.88	7.95	12.83	17.83	22.49	24.87	25.89	22.26	16.92	10.68	5.56	4.09	17.76
West Virginia													
Charleston	7.04	9.65	13.40	17.15	20.21	21.69	20.90	18.97	15.56	11.81	7.72	6.02	14.20
Elkins	6.93	9.43	12.83	16.35	19.08	20.56	19.99	18.06	14.88	11.13	7.27	5.79	13.51
Wisconsin													
Green Bay	6.25	9.31	13.17	16.81	20.56	22.49	22.03	18.85	14.20	9.65	5.79	4.88	13.74
Madison	6.59	9.88	13.29	16.92	20.67	22.83	22.37	19.42	14.76	3.41	6.25	5.22	14.08
Milwaukee	6.47	9.31	12.72	16.69	20.78	22.94	22.60	19.42	14.88	10.22	6.25	5.11	13.97
Wyoming													
Rock Springs	7.61	10.90	15.10	19.42	23.17	26.01	25.78	22.94	18.62	13.40	8.40	6.70	16.58
Seridan	6.47	9.77	13.97	17.94	20.90	23.85	24.64	21.69	16.47	11.24	7.15	5.56	14.99

(Source: National Renewable Energy Laboratory, USA; Internet Address: http://rredc.nrel.gov/solar)

Solar Radiation on Tilted Surfaces

The worldwide horizontal surface data on pages 18-31 can be used to estimate solar radiation on tilted surfaces. This section outlines a procedure to use this data to estimate the solar radiation on equatorial-facing, tilted collectors. Nomographs are provided for common tilt angles of: tilt equal to the site latitude and the latitude +/- 15°.

Procedure:

1. **Obtain the monthly averaged, daily total radiation, $\overline{H}_h$,** for the chosen location and month of interest, pages 18-31.

2. **Compute the monthly clearness index, $\overline{K}_T$:**

$$\overline{K}_T = \frac{\overline{H}_h}{\overline{H}_{o,h}}$$ Where $\overline{H}_{o,h}$ is the extraterrestrial horizontal

surface radiation and can be read from Fig. c-1 on page 33.

3. **Determine the sunset hour angle,** h_{ss}, from Fig. c-2 page 34, (positive value).

4. **Compute the diffuse to total radiation ratio, $\dfrac{\overline{D}_h}{\overline{H}_h}$:**

$$\frac{\overline{D}_h}{\overline{H}_h} = 0.775 + 0.347\left(h_{ss} - \frac{\pi}{2}\right) - \left[0.505 + 0.0261\left(h_{ss} - \frac{\pi}{2}\right)\right]\cos\left(2\overline{K}_T - 1.8\right)$$

Note: h_{ss} in radians.

5. **Solve for the diffuse and beam radiation components:**

$$\overline{D}_h = \left(\frac{\overline{D}_h}{\overline{H}_h}\right)\overline{H}_h \qquad \text{and} \qquad \overline{B}_h = \overline{H}_h - \overline{D}_h$$

6. **Determine the collector tilt factor, $\overline{R}_b$, for the tilt angle, β, of interest.** Use the following formulation or Figs. c-3-5 for common tilt orientations.

$$\overline{R}_b = \frac{\cos(L-\beta)\cos(\delta_s)\sin(h_{sr}) + h_{sr}\sin(L-\beta)\sin(\delta_s)}{\cos(L)\cos(\delta_s)\sin(h_{sr}) + h_{sr}\sin(L)\sin(\delta_s)}$$

Where δ_s is the declination angle, page 4, and h_{sr} is the hour angle at sunrise, Fig. c-2 (negative value, radians).

7. **Compute the collector monthly-averaged radiation total, $\overline{H}_c$:**

$$\overline{H}_c = \overline{R}_b\overline{B}_h + \overline{D}_h\cos^2\left(\frac{\beta}{2}\right) + (\overline{D}_h + \overline{B}_h)\rho\sin^2\left(\frac{\beta}{2}\right)$$

Assuming an appropriate reflectivity value, ρ, from the table on page 36.

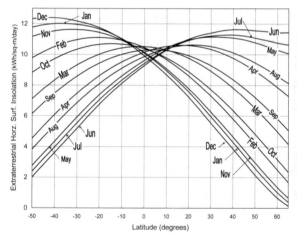

Figure c-1. $\overline{H}_{o,h}$, extraterrestrial ,monthly averaged, daily insolation on a horizontal surface.

33

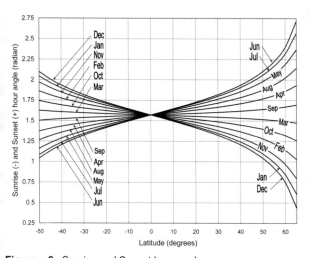

Figure c-2. Sunrise and Sunset hour angles.

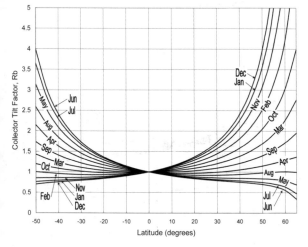

Figure c-3. $\overline{R}_b$ for tilt = L.

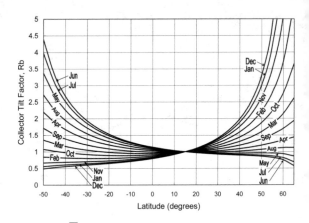

Figure c-4. $\overline{R}_b$ for tilt = L – 15.

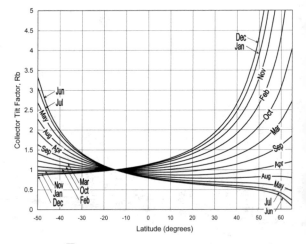

Figure c-5. $\overline{R}_b$ for tilt = L + 15.

Reflectivity values for characteristic surfaces (integrated over solar spectrum and angle of incidence)[a].

Surface	Average reflectivity
Snow (freshly fallen or with ice film)	0.75
Water surfaces (relatively large incidence angles)	0.07
Soils (clay, loam, etc.)	0.14
Earth roads	0.04
Coniferous forest (winter)	0.07
Forests in autumn, ripe field crops, plants	0.26
Weathered blacktop	0.10
Weathered concrete	0.22
Dead leaves	0.30
Dry grass	0.20
Green grass	0.26
Bituminous and gravel roof	0.13
Crushed rock surface	0.20
Building surfaces, dark (red brick, dark paints, etc.)	0.27
Building surfaces, light (light brick, light paints, etc.)	0.60

[a]From Hunn, B. D., and D. O. Calafell, Determination of Average Ground Reflectivity for Solar Collectors, *Sol. Energy*, vol. 19, p. 87, 1977; see also R. J. List, "Smithsonian Meteorological Tables," 6th ed., Smithsonian Institution Press, pp. 442–443, 1949.

Estimation of UV Insolation

Nomenclature

I_t	global horizontal total solar radiation
$I_{t,b}$	beam component of total solar radiation
$I_{UV,b}$	beam component of UV radiation
$I_{UV,h}$	global horizontal UV radiation
K_t	cloudiness index
m	air mass
α	solar altitude angle

$$\frac{I_{UV,h}}{I_t} = 0.14315 K_t^2 - 0.20445 K_t + 0.135544$$

$$\frac{I_{UV,b}}{I_{t,b}} = 0.0688 e^{-0.575m} = 0.688 \exp\left(\frac{-0.575}{\sin \alpha}\right)$$

Radiation Heat Transfer Properties

Surface emissivity, ε, which is the emission of the surface compared to an ideal blackbody, E/E_b.

$$\varepsilon \equiv \frac{E}{E_b} = \frac{1}{\sigma T^4} \int_0^\infty \varepsilon_\lambda E_{b\lambda} \, d\lambda$$

Where σ is the Stefan-Boltzmann constant, (5.670×10^{-8} W/m²-K⁴), T is surface temperature and the subscript λ refers to wavelength.

Incident radiation balance: $\alpha + \tau + \rho = 1$

Where:

absorptivity: $\alpha \equiv \dfrac{I_{absorbed}}{I}$ transmissivity: $\tau \equiv \dfrac{I_{trans.}}{I}$

reflectivity: $\rho \equiv \dfrac{I_{refl.}}{I}$ and I is the total surface irradiation.

Emissivity and absorptivity of common materials

Substance	Short-wave absorptance	Long-wave emittance	α / ε
Class I substances: Absorptance to emittance ratios less than 0.5			
Magnesium carbonate, MgCO₃	0.025–0.04	0.79	0.03–0.05
White plaster	0.07	0.91	0.08
Snow, fine particles, fresh	0.13	0.82	0.16
White paint, 0.017 in, on aluminum	0.20	0.91	0.22
Whitewash on galvanized iron	0.22	0.90	0.24
White paper	0.25–0.28	0.95	0.26–0.29
White enamel on iron	0.25–0.45	0.9	0.28–0.5
Ice, with sparse snow cover	0.31	0.96–0.97	0.32
Snow, ice granules	0.33	0.89	0.37
Aluminum oil base paint	0.45	0.90	0.50
White powdered sand	0.45	0.84	0.54

Substance	Short-wave absorptance	Long-wave emittance	α / ε
Class II substances: Absorptance to emittance ratios between 0.5 and 0.9			
Asbestos felt	0.25	0.50	0.50
Green oil base paint	0.5	0.9	0.56
Bricks, red	0.55	0.92	0.60
Asbestos cement board, white	0.59	0.96	0.61
Marble, polished	0.5–0.6	0.9	0.61
Wood, planed oak	—	0.9	—
Rough concrete	0.60	0.97	0.62
Concrete	0.60	0.88	0.68
Grass, green, after rain	0.67	0.98	0.68
Grass, high and dry	0.67–0.69	0.9	0.76
Vegetable fields and shrubs, wilted	0.70	0.9	0.78
Oak leaves	0.71–0.78	0.91–0.95	0.78–0.82
Frozen soil	—	0.93–0.94	—
Desert surface	0.75	0.9	0.83
Common vegetable fields and shrubs	0.72–0.76	0.9	0.82
Ground, dry plowed	0.75–0.80	0.9	0.83–0.89
Oak woodland	0.82	0.9	0.91
Pine forest	0.86	0.9	0.96
Earth surface as a whole (land and sea, no clouds)	0.83	—	—
Class III substances: Absorptance to emittance ratios between 0.8 and 1.0			
Grey paint	0.75	0.95	0.79
Red oil base paint	0.74	0.90	0.82
Asbestos, slate	0.81	0.96	0.84
Asbestos, paper		0.93–0.96	—
Linoleum, red-brown	0.84	0.92	0.91
Dry sand	0.82	0.90	0.91
Green roll roofing	0.88	0.91–0.97	0.93
Slate, dark grey	0.89	—	—
Old grey rubber	—	0.86	—
Hard black rubber	—	0.90–0.95	—
Asphalt pavement	0.93	—	—
Black cupric oxide on copper	0.91	0.96	0.95
Bare moist ground	0.9	0.95	0.95
Wet sand	0.91	0.95	0.96
Water	0.94	0.95–0.96	0.98
Black tar paper	0.93	0.93	1.0
Black gloss paint	0.90	0.90	1.0
Small hole in large box, furnace, or enclosure	0.99	0.99	1.0
"Hohlraum," theoretically perfect black body	1.0	1.0	1.0

Emissivity and absorptivity of common materials,

continued from page 37.

Substance	Short-wave absorptance	Long-wave emittance	$\frac{\alpha}{\varepsilon}$
Class IV substances: Absorptance to emittance ratios greater than 1.0			
Black silk velvet	0.99	0.97	1.02
Alfalfa, dark green	0.97	0.95	1.02
Lampblack	0.98	0.95	1.03
Black paint, 0.017 in, on aluminum	0.94–0.98	0.88	1.07–1.11
Granite	0.55	0.44	1.25
Graphite	0.78	0.41	1.90
High ratios, but absorptances less than 0.80			
Dull brass, copper, lead	0.2–0.4	0.4–0.65	1.63–2.0
Galvanized sheet iron, oxidized	0.8	0.28	2.86
Galvanized iron, clean, new	0.65	0.13	5.0
Aluminum foil	0.15	0.05	3.00
Magnesium	0.3	0.07	4.3
Chromium	0.49	0.08	6.13
Polished zinc	0.46	0.02	23.0
Deposited silver (optical reflector) untarnished	0.07	0.01	
Class V substances: Selective surfaces[b]			
Plated metals:[c]			
Black sulfide on metal	0.92	0.10	9.2
Black cupric oxide on sheet aluminum	0.08–0.93	0.09–0.21	
Copper (5×10^{-5} cm thick) on nickel or silver-plated metal			
Cobalt oxide on platinum			
Cobalt oxide on polished nickel	0.93–0.94	0.24–0.40	3.9
Black nickel oxide on aluminum	0.85–0.93	0.06–0.1	14.5–15.5
Black chrome	0.87	0.09	9.8
Particulate coatings:			
Lampblack on metal			
Black iron oxide, 47 μm grain size, on aluminum			
Geometrically enhanced surfaces:[d]			
Optimally corrugated greys	0.89	0.77	1.2
Optimally corrugated selectives	0.95	0.16	5.9
Stainless-steel wire mesh	0.63–0.86	0.23–0.28	2.7–3.0
Copper, treated with $NaClO_2$ and NaOH	0.87	0.13	6.69

[a]From Anderson, B., "Solar Energy," McGraw-Hill Book Company, 1977, with permission.

[b]Selective surfaces absorb most of the solar radiation between 0.3 and 1.9 μm, and emit very little in the 5–15 μm range—the infrared.

[c]For a discussion of plated selective surfaces, see Daniels, "Direct Use of the Sun's Energy," especially chapter 12.

[d]For a discussion of how surface selectivity can be enhanced through surface geometry, see K. G. T. Hollands, Directional Selectivity Emittance and Absorptance Properties of Vee Corrugated Specular Surfaces, J. Sol. Energy Sci. Eng., vol. 3, July 1963.

Properties of some selective plated coating systems[a]

Coating[b]	Substrate	$\bar{\alpha}_s$	$\bar{\epsilon}_i$	Breakdown temperature (°C)	Durability — Humidity-Degradation MIL. STD 810B
Black nickel on nickel	Steel	0.95	0.07	>290	Variable
Black chrome on nickel	Steel	0.95	0.09	>430	No effect
Black chrome	Steel	0.91	0.07	>430	Completely rusted
	Copper	0.95	0.14	315	Little effect
	Galvanized steel	0.95	0.16	>430	Complete removal
Black copper	Copper	0.88	0.15	315	Complete removal
Iron oxide	Steel	0.85	0.08	430	Little effect
Manganese oxide	Aluminum	0.70	0.08		
Organic overcoat on iron oxide	Steel	0.90	0.16		Little effect
Organic overcoat on black chrome	Steel	0.94	0.20		Little effect

[a] From U.S. Dept. of Commerce, "Optical Coatings for Flat Plate Solar Collectors," NTIS No. PN-252-383, Honeywell, Inc., 1975.

[b] Black nickel coating plated over a nickel-steel substrate has the best selective properties ($\bar{\alpha}_s = 0.95$, $\bar{\epsilon}_i = 0.07$), but these degraded significantly during humidity tests. Black chrome plated on a nickel-steel substrate also had very good selective properties ($\bar{\alpha}_s = 0.95$, $\bar{\epsilon}_i = 0.09$) and also showed high resistance to humidity.

Angular variation of the absorptivity of lampblack paint

Incidence angle $i(°)$	Absorptance $\alpha(i)$
0–30	0.96
30–40	0.95
40–50	0.93
50–60	0.91
60–70	0.88
70–80	0.81
80–90	0.66

Adapted from [9]

Reflectivity values for reflector materials

Material	ρ
Silver (unstable as front surface mirror)	0.94 ± 0.02
Gold	0.76 ± 0.03
Aluminized acrylic, second surface	0.86
Anodized aluminum	0.82 ± 0.05
Various aluminum surfaces-range	0.82–0.92
Copper	0.75
Back-silvered water-white plate glass	0.88
Aluminized type-C Mylar (from Mylar side)	0.76

Spectral Absorption of Solar Radiation in Water

	Layer depth				
Wavelength (μm)	0	1 cm	10 cm	1 m	10 m
0.2–0.6	23.7	23.7	23.6	22.9	17.2
0.6–0.9	36.0	35.3	36.0	12.9	0.9
0.9–1.2	17.9	12.3	0.8	0.0	0.0
1.2 and over	22.4	1.7	0.0	0.0	0.0
Total	100.0	73.0	54.9	35.8	18.1

[a]Numbers in the table give the percentage of sunlight in the wavelength band passing through water of the indicated thickness.

Transparent Materials

Refraction of light between materials

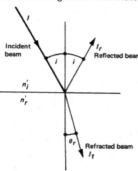

Index of refraction:

$$\frac{\sin(i)}{\sin(\theta_r)} = \frac{n'_r}{n'_i} = n_r$$

Visible spectrum refractive index values, n_r, based on air

Material	Index of Refraction
Air	1.000
Clean polycarbonate	1.59
Diamond	2.42
Glass (solar collector type)	1.50–1.52
Plexiglass[a] (polymethyl methacrylate, PMMA)	1.49
Mylar[a] (polyethylene terephthalate, PET)	1.64
Quartz	1.54
Tedlar[a] (polyvinyl fluoride, PVF)	1.45
Teflon[a] (polyfluoroethylenepropylene, FEP)	1.34
Water–liquid	1.33
Water–solid	1.31

[a]Trademark of the duPont Company, Wilmington, Delaware.

Thermal and radiative properties of collector cover materials[a]

Material name	Index of refraction (n)	τ (solar)[b] (%)	τ (solar)[b] (%)	τ (infrared)[b] (%)	Expansion coefficient (m/m × °C)	Temperature limits (°C)	Weatherability (comment)	Chemical resistance (comment)
Lexan (polycarbonate)	1.586 (D 542)γ	125 mil 64.1 (±0.8)	125 mil 72.6 (±0.1)	125 mil 2.0 (est)[d]	6.75 (10⁻⁵) (H 696)	121-132 service temperature	Good: 2 yr exposure in Florida caused yellowing; 5 yr caused 5% loss in τ	Good: comparable to acrylic
Plexiglas (acrylic)	1.49 (D 542)	125 mil 89.6 (±0.3)	125 mil 79.6 (±0.8)	125 mil 2.0 (est)[d]	7.02 (10⁻⁵) at 15.5°C; 8.28 (10⁻⁵) at 38°C	82-93 service temperature	Average to good: based on 20 yr testing in Arizona, Florida, and Pennsylvania	Good to excellent: resists most acids and alkalis
Teflon F.E.P. (fluorocarbon)	1.343 (D 542)	5 mil 92.3 (±0.2)	5 mil 89.8 (±0.4)	5 mil 25.6 (±0.5)	1.06(10⁻⁴) at 71°C; 1.6 (10⁻⁴) at 100°C	204 continuous use; 246 short-term use	Good to excellent: based on 15 yr exposure in Florida environment	Excellent: chemically inert
Tedlar P.V.F. (fluorocarbon)	1.46 (D 542)	4 mil 92.2 (±0.1)	4 mil 88.3 (±0.9)	4 mil 20.7 (±0.2)	5.04 (10⁻⁵) (D 696)	107 continuous use; 177 short-term use	Good to excellent: 10 yr exposure in Florida with slight yellowing	Excellent: chemically inert
Mylar (polyester)	1.64-1.67 (D 542)	5 mil 86.9 (±0.3)	5 mil 80.1 (±0.1)	5 mil 17.8 (±0.5)	1.69 (10⁻⁵) (D 696-44)	150 continuous use; 204 short-term use	Poor: ultraviolet degradation great	Good to excellent: comparable to Tedlar
Sunlite/(fiberglass)	1.54 (D 542)	25 mil (P) 86.5 (±0.2) 25 mil (R) 87.5 (±0.2)	25 mil (P) 75.4 (±0.1) 25 mil (R) 77.1 (±0.7)	25 mil (P) 7.6 (±0.1) 25 mil (R) 3.3 (±0.3)	2.5 (10⁻⁵) (D 696)	93 continuous use; causes 5% loss in τ	Fair to good: regular, 7 yr solar life; premium, 20 yr solar life	Good: inert to chemical atmospheres
Float glass (glass)	1.518 (D 542)	125 mil 84.3 (±0.1)	125 mil 78.6 (±0.2)	125 mil 2.0 (est)[d]	8.64 (10⁻⁵) (D 696)	732 softening point; 38 thermal shock	Excellent: time proved	Good to excellent: time proved

Thermal and radiative properties of collector cover materials[a]

Material name	Index of refraction (n)	τ (solar)[a] (%)	τ (solar)[f] (%)	τ (infrared)[b] (%)	Expansion coefficient (m/m × °C)	Temperature limits (°C)	Weatherability (comment)	Chemical resistance (comment)
Temper glass (glass)	1.518 (D 542)	125 mil 84.3 (±0.1)	125 mil 78.6 (±0.2)	125 mil 2.0	8.64 (10-6) (D696)	232–260 continuous use; 260–288 short-term use	Excellent: time proved	Good to excellent: time proved
Clear lime sheet glass (low iron oxide glass)	1.51 (D 542)	Insufficient data provided by ASG	125 mil 87.5 (±0.5)	125 mil 2.0 (est)[d]	9 (10-6) (D 696)	204 for continuous operation	Excellent: time proved	Good to excellent: time proved
Clear lime temper glass (low iron oxide glass)	1.51 (D 542)	Insufficient data provided by ASG	125 mil 87.5 (±0.5)	125 mil 2.0 (est)	9 (10-6) (D 696)	204 for continuous operation	Excellent: time proved	Good to excellent: time proved
Sunadex white crystal glass (0.01% iron oxide glass)	1.50 (D 542)	Insufficient data provided by ASG	125 mil 91.5 (±0.2)	125 mil 2.0 (est)	8.46 (10-6) (D 696)	204 for continuous operation	Excellent: time proved	Good to excellent: time proved

[a] Numerical integration ($\Sigma \tau_{swg} F_{\lambda, T-t_{\lambda,T}}$) for $\lambda = 0.2$–4.0 μM.

[b] Numerical integration ($\Sigma \tau_{swg} F_{\lambda, T-t_{\lambda,T}}$) for $\lambda = 3.0$–50.0 μM.

[c] All parenthesized numbers refer to ASTM test codes.

[d] Data not provided; estimate of 2% to be used for 125 mil samples.

[e] Degrees differential to rupture 2 × 2 × ¼ in samples. Glass specimens heated and then quenched in water bath at 70°F.

[f] Sunlite premium data denoted by (P); Sunlite regular data denoted by (R).

[g] Compiled data based on ASTM Code E 424 Method B.

[h] Abstracted from Ratzel, A. C., and R. B. Bannerot, Optimal Material Selection for Flat-Plate Solar Energy Collectors Utilizing Commercially Available Materials, presented at ASME–AIChE Natl. Heat Transfer Conf., 1976.

Solar Thermal Collector Overview

General Configuration	Description	Concentration Ratio	Indicative Operating Temp. (°C)
	Non-Convecting Solar Pond	1	30-70
	Unglazed Flat Plate Absorber	1	O 40
	Flat Plate Collector (High Efficiency)	1 (1)	O 70 (60-120)
	Fixed Concentrator	3-5	100-150
	Evacuated Tube	1	50-180
	Compound Parabolic (With 1 Axis Tracking)	1-5 (5-15)	70-240 (70-290)
	Parabolic Trough	10-50	150-350
	Fresnel Refractor	10-40	70-270
	Spherical Dish Reflector	100-300	70-730
	Parabolic Dish Reflector	200-500	250-700
	Central Receiver	500-3000	500->1000

Adapted from [12]

45

Non-Concentrating Collectors

Efficiency of N-C collectors, Hottel-Whillier equation (European standard on page 48):

$$\eta_c = F_R \tau_s \alpha_s - F_R U_c \frac{(T_{f,in} - T_a)}{I_c}$$

Where F_R is the collector heat removal factor, τ_s is the cover transmissivity, α_s is the cover-absorber absorptivity, and U_c is the overall collector heat loss conductance. $T_{f,in}$ is the collector fluid inlet temperature, T_a is the ambient, and I_c is the incident radiation on the collector.

Representative performance curves

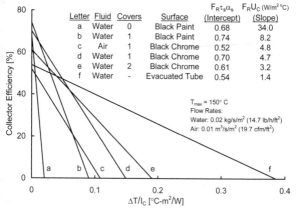

Letter	Fluid	Covers	Surface	$F_R \tau_s \alpha_s$ (Intercept)	$F_R U_C$ (W/m² °C) (Slope)
a	Water	0	Black Paint	0.68	34.0
b	Water	1	Black Paint	0.74	8.2
c	Air	1	Black Chrome	0.52	4.8
d	Water	1	Black Chrome	0.70	4.7
e	Water	2	Black Chrome	0.61	3.2
f	Water	-	Evacuated Tube	0.54	1.4

$T_{max} = 150°$ C
Flow Rates:
Water: 0.02 kg/s/m² (14.7 lb/h/ft²)
Air: 0.01 m³/s/m² (19.7 cfm/ft²)

Incidence Angle Modifier

Incidence angle modifier, $K_{\tau\alpha}$, is used to estimate collector performance at non-normal angles of incidence with:

$$\eta_c = F_R \left[K_{\tau\alpha} (\tau_s \alpha_s)_n - U_c \frac{(T_{f,in} - T_a)}{I_c} \right]$$

where $K_{\tau\alpha}$ is of the form (b = constant):

$$K_{\tau\alpha} = 1 - b\left(\frac{1}{\cos(i)} - 1\right)$$

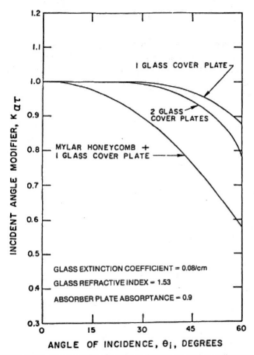

Incident angle modifier for three flat-plate solar collectors.
Reprinted by permission of the American Society of Heating,
Refrigerating and Air-Conditioning Engineers, Inc., Atlanta, from
ASHRAE Standard 93-77, "Methods of Testing to Determine the
Thermal Performance of Solar Collectors."

47

European Standard N-C collector efficiency formulation:

$$\eta_c = F_R \tau_s \alpha_s - a_1 \frac{T_{ave} - T_{amb}}{G} - a_2 \frac{\left(T_{ave} - T_{amb}\right)^2}{G}$$

Where $T_{ave} = 0.5(T_{f,in} + T_{f,out})$, is used instead of collector inlet temperature and G is the hemispherical irradiance (W/m^2).

The incidence angle modifier is applied in a similar manner:

$$\tau_s \alpha_s = K_{\tau\alpha} \left(\tau_s \alpha_s\right)_n$$

Concentrating Thermal Collectors

Geometric Concentration Ratio, CR:

$$CR = \frac{A_a}{A_r} \qquad \text{where } A_a \text{ is the aperture area and } A_r \text{ is the}$$

receiver area.

Instantaneous collector efficiency, η_c :

$$\eta_c = \eta_o - U_c \frac{\left(T_r - T_a\right)}{I_c \left(CR\right)}$$

Where η_o is the optical efficiency, U_c is the overall collector heat loss conductance, T_r is the receiver temperature, T_a is the ambient temperature, and I_c is the incident radiation on the collector.

Incidence factors for various orientation and tracking arrangements of concentrating collectors

Orientation of collector	Incidence factor $\cos i$
Fixed, horizontal, plane surface.	$\sin L \sin \delta_s + \cos \delta_s \cos h_s \cos L$
Fixed plane surface tilted so that it is normal to the solar beam at noon on the equinoxes.	$\cos \delta_s \cos h_s$
Rotation of a plane surface about a horizontal east-west axis with a single daily adjustment permitted so that its surface normal coincides with the solar beam at noon every day of the year.	$\sin^2 \delta_s + \cos^2 \delta_s \cos h_s$
Rotation of a plane surface about a horizontal east-west axis with continuous adjustment to obtain maximum energy incidence.	$\sqrt{1 - \cos^2 \delta_s \sin^2 h_s}$
Rotation of a plane surface about a horizontal north-south axis with continuous adjustment to obtain maximum energy incidence.	$[(\sin L \sin \delta_s + \cos L \cos \delta_s \cos h_s)^2 + \cos^2 \delta_s \sin^2 h_s]^{1/2}$
Rotation of a plane surface about an axis parallel to the earth's axis with continuous adjustment to obtain maximum energy incidence.	$\cos \delta_s$
Rotation about two perpendicular axes with continuous adjustment to allow the surface normal to coincide with the solar beam at all times.	1

^{a}The incidence factor denotes the cosine of the angle between the surface normal and the solar beam.

Adaptation of monthly-averaged, horizontal surface data for tracking concentrators:

$$\overline{H}_c = \left[\overline{r}_T - \overline{r}_d \left(\overline{D}_h \middle/ \overline{H}_h \right) \right] \overline{H}_h \quad \text{where:}$$

$\overline{D}_h$ monthly-average diffuse radiation component for a horizontal surface

$\overline{H}_c$ monthly-average terrestrial radiation for a tracking collector

$\overline{H}_h$ monthly-average terrestrial radiation for a horizontal surface

$\overline{r}_d$ diffuse radiation factor

$\overline{r}_T$ tracking factor

49

Thermal Energy Storage

Sensible heat storage:
$$Q_{sensible} = \rho V c_p \Delta T$$

Combined latent and sensible heat storage:

$$Q_{total} = m \left[\overline{c}_{p_{solid}} \left(T_{melt} - T_{low} \right) + \lambda + \overline{c}_{p_{liquid}} \left(T_{high} - T_{melt} \right) \right]$$

Thermochemical energy storage:

$$Q_{thermochemical} = a_r m \Delta H$$

Where m is the mass of reactant, a_r is the fraction reacted and ΔH is the heat of reaction per unit mass.

Storage materials properties given on pages 51-53.

Thermal conductivities of containment materials

Materials	Thermal conductivity[b]	
	(W/m K)	(Btu in/hr ft² °F)
Plastics[a]		
ABS	0.17–0.33	1.2–2.3
Acrylic	0.19–0.43	1.3–3.0
Polypropylene	0.12–0.17	0.8–1.2
Polyethylene (high density)	0.43–0.52	3.0–3.6
Polyethylene (medium density)	0.30–0.42	2.1–2.9
Polyethylene (low density)	0.30	2.1
Polyvinyl chloride	0.13	0.9
Metals[c]		
Aluminum	200	1500
Copper	390	2700
Steel	48	330

[a]*Plastics, a desk-top data bank*, 5th Ed. Book A. San Diego, CA: The International Plastics Selector, Inc., 1980.
[b]As measured by ASTM C-177.
[c]*Handbook of Chemistry and Physics*, 40th Ed.

Physical properties of some sensible heat storage materials

Storage Medium	Temperature Range, °C	Density (ρ), kg/m³	Specific Heat (C), J/kg K	Energy Density (ρC) kWh/m³ K	Thermal Conductivity (W/m K)
Water	0–100	1000	4190	1.16	0.63 at 38°C
Water (10 bar)	0–180	881	4190	1.03	—
50% ethylene glycol–50% water	0–100	1075	3480	0.98	—
Dowtherm A® (Dow Chemical, Co.)	12–260	867	2200	0.53	0.122 at 260°C
Therminol 66® (Monsanto Co.)	–9–343	750	2100	0.44	0.106 at 343°C
Draw salt (50NaNO₃–50KNO₃)ᵃ	220–540	1733	1550	0.75	0.57
Molten salt (53KNO₃/40NaNO₂/7NaNO₃)ᵃ	142–540	1680	1560	0.72	0.61
Liquid Sodium	100–760	750	1260	0.26	67.5
Cast iron	m.p. (1150–1300)	7200	540	1.08	42.0
Taconite	—	3200	800	0.71	—
Aluminum	m.p. 660	2700	920	0.69	200
Fireclay	—	2100–2600	1000	0.65	1.0–1.5
Rock	—	1600	880	0.39	—

ᵃ Composition in percent by weight.
Note: m.p. = melting point.

Physical properties of latent heat storage materials or PCMs

Storage Medium	Melting Point °C	Latent Heat, kJ/kg	Specific Heat (kJ/kg °C)		Density (Kg/m³)		Energy Density (kWhr/m³K)	Thermal Conductivity (W/m K)
			Solid	Liquid	Solid	Liquid		
$LiClO_3 \cdot 3H_2O$	8.1	253	—	—	1720	1530	108	—
$Na_2SO_4 \cdot 10H_2O$ (Glauber's Salt)	32.4	251	1.76	3.32	1460	1330	92.7	2.25
$Na_2S_2O_3 \cdot 5H_2O$	48	200	1.47	2.39	1730	1665	92.5	0.57
$NaCH_3COO \cdot 3H_2O$	58	180	1.90	2.50	1450	1280	64	0.5
$Ba(OH)_2 \cdot 8H_2O$	78	301	0.67	1.26	2070	1937	162	0.653ℓ
$Mg(NO_3)_2 \cdot 6H_2O$	90	163	1.56	3.68	1636	1550	70	0.611
$LiNO_3$	252	530	2.02	2.041	2310	1776	261	1.35
$LiCO_3/K_2CO_3$ (35:65)ᵃ	505	345	1.34	1.76	2265	1960	188	—
$LiCO_3/K_2CO_3/Na_2CO_3$ (32:35:33)ᵃ	397	277	1.68	1.63	2300	2140	165	—
n-Tetradecane	5.5	228	—	—	825	771	48	0.150
n-Octadecane	28	244	2.16	—	814	774	52.5	0.150
HDPE (cross-linked)	126	180	2.88	2.51	960	900	45	0.361
Steric acid	70	203	—	2.35	941	347	48	0.172ℓ

ᵃComposition in percent by weight.
Note: ℓ = liquid.

Properties of thermochemical storage media

Reaction	Condition of Reaction		Component (Phase)	Temperature, °C	Pressure, kPa	Density, kg/m³	Volumetric Storage Density, kWh/m³
	Pressure, kPa	Temperature, °C					
$MgCO_3(s) + 1200$ kJ/kg = $MgO(s) + CO_2(g)$	100	427–327	$MgCO_3(s)$ $CO_2(\ell)$	20 31	100 7400	1500 465	187
$Ca(OH)_2(s) + 1415$ kJ/kg = $CaO(s) + H_2O(g)$	100	572–402	$Ca(OH)_2(s)$ $H_2O(\ell)$	20	100	1115	345
$SO_3(g) + 1235$ kJ/kg = $SO_2(g) + \frac{1}{2}O_2(g)$	100	520–960	$SO_3(\ell)$ $SO_2(\ell)$ $O_2(g)$	45 40 20	100 630 10000	1900 1320 130	280

Note: s = solid; ℓ = liquid; g = gas

Photovoltaics
Cell Output Characteristics

Cell output power across load, P_L: $\qquad P_L = I_L V = I_L^2 R_L$

Where I_L is the load circuit current, V is the voltage and R_L is the load resistance.

Representative current, voltage, and power outputs of a PV cell

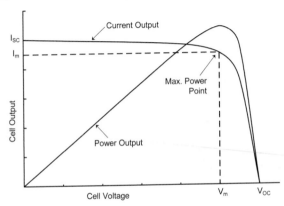

Open circuit junction voltage: $\quad V_{oc} = \dfrac{kT}{e_o} \ln\left(\dfrac{I_{sc}}{I_o} + 1\right)$

Where:
k	Boltzman's constant, $(1.381 \times 10^{-23}$ J/K)
T	cell temperature, (K)
e_o	electron charge, $(1.602 \times 10^{-19}$ J/V)
I_{sc}	short circuit current
I_o	reverse saturation current

Voltage at maximum power output, V_m, is found from:

$$\exp\left(\dfrac{e_o V_m}{kT}\right)\left(1 + \dfrac{e_o V_m}{kT}\right) = 1 + \dfrac{I_{sc}}{I_o}$$

54

Effect of illumination and load resistance on PV cell output

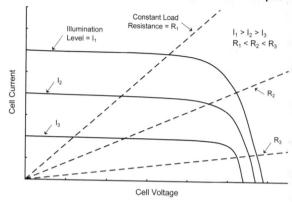

Effect of temperature on cell output

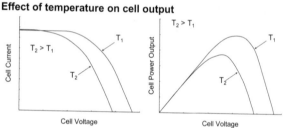

Temperature corrections for the cell output are supplied by the manufacturers and are typically of the form:

$$V = V_{ref}\left(1 + \beta\left(T_c - T_{ref}\right)\right) \quad \text{and} \quad I = I_{ref}\left(1 + \alpha\left(T_c - T_{ref}\right)\right)$$

where the subscript "ref" refers to values at a reference condition and T_c is the actual operating temperature of the cell. α and β are constants provided by the manufacturer. Since α is much smaller than β, power output goes down as the temperature of the cell goes up.

Multiple cell output

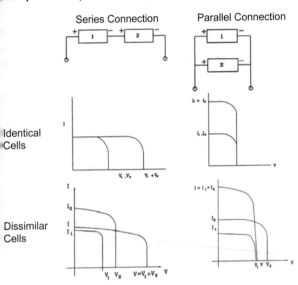

Series Connection   Parallel Connection

Identical
Cells

Dissimilar
Cells

Effect of collector tilt and tracking on incident solar radiation

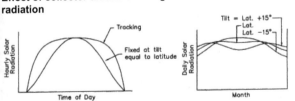

PV Power Configurations

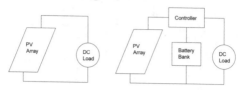

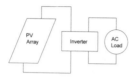

Direct PV-load connection, without and with storage

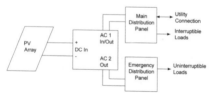

AC power supply

Utility interconnected configuration

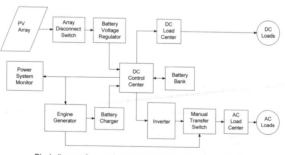

Block diagram for stand-alone PV system with generator backup

Adapted from [10,13]

PV System Design

System Design Suggestions from [13]

Keep it simple – Complexity lowers reliability and increases maintenance costs.

Understand system availability – Achieving 99+ % availability with <u>any</u> energy system is expensive.

Be thorough, but realistic, when estimating the load – A large safety factor can cost you a great deal of money.

Cross-check weather sources – Errors in solar resource estimates can cause disappointing system performance.

Know what hardware is available at what cost – Tradeoffs are inevitable. The more you know about hardware, the better decisions you can make. Shop for bargains, talk to dealers, ask questions.

Install the system carefully – Make each connection as if it had to last 30 years—it does. Use the right tools and technique. The system reliability is no higher than its weakest connection.

Safety first and last – Don't take shortcuts that might endanger life or property. Comply with local and national building and electrical codes.

Plan periodic maintenance – PV systems have an enviable record for unattended operation, but no system works forever without some care.

Calculate the life-cycle cost (LCC) to compare PV systems to alternatives – LCC reflects the complete cost of owning and operating any energy system.

Simplified Average Daily Load Determination from [10]

1. Identify all loads to be connected to the PV system.
2. For each load, determine its voltage, current, power and daily operating hours. For some loads, the operation may vary on a daily, monthly or seasonal basis. If so, this must be accounted for in calculating daily averages.
3. Separate ac loads from dc loads.
4. Determine average daily Ah for each load from current and operating hours data. If operating hours differ from day to day during the week, the daily average over the week should be calculated. If average daily operating hours vary

from month to month, then the load calculation may need to be determined for each month.

5. Add up the Ah for the dc loads, being sure all are at the same voltage.

6. If some dc loads are at a different voltage, which will require a dc to dc converter, then the converter input Ah for these loads needs to account for the conversion efficiency of the converter.

7. For ac loads, the dc input current to the inverter must be determined and the dc Ah are then determined from the dc input current. The dc input current is determined by equating the ac load power to the dc input power and then dividing by the efficiency of the inverter.

8. Add the Ah for the dc loads to the Ah for the ac loads, then divide by the wire efficiency factor and the battery efficiency factor to obtain the corrected average daily Ah for the total load.

9. The total ac power will determine the required size of the inverter. Individual load powers will be needed to determine wire sizing to the loads. Total load current will be compared with total array current when sizing wire from battery to controller.

Storage Estimation: Estimation of the days of battery storage needed for a stand-alone system if no better estimate is available, [10].

$$D_{crit} = -1.9T_{min} + 18.3 \qquad \text{or}$$

$$D_{non} = -0.48T_{min} + 4.58$$

Where:

D_{crit} number of storage days required for critical application
D_{non} number of storage days required for non-critical application
T_{min} minimum average daily peak sun hours for selected collector tilt during any month of operation

Note: $T_{min} \geq 1$ peak sun hours per day

Water Pumping Load Data

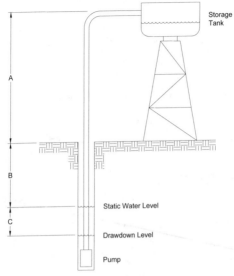

Pumping power (W):

$$P = \frac{\rho_w \dot{V} g H}{\eta_p}$$

Where:

g   gravitational acceleration, (9.81 m/s²)

H   total pumping head, (m)

H_d dynamic head, (m)

H_f friction head, (m)

H_s static head, (m)

$\dot{V}$ volumetric flow rate, (m³/s)

η_p pump efficiency

v   water velocity at pipe outlet, (m/s)

ρ_w water density, (997 kg/m³)

And: $H = H_d + H_s$ $H_d = H_f + \dfrac{v^2}{2g}$

$$H_s = \begin{cases} A + B, \text{ for no draw down} \\ A + B + C, \text{ for water level drawn down a depth } C \end{cases}$$

Representative water pump operating characteristics

Head (m)	Type pump	Wire-to-water efficiency (%)
0–5	Centrifugal	15–25
6–20	Centrifugal with Jet	10–20
	Submersible	20–30
21–100	Submersible	30–40
	Jack pump	30–45
>100	Jack pump	35–50

Electrical wire load rating

Resistance and amperage ratings for type THHN insulated wire.

AWG Wire Size	Resistance @ 20°C (Ω/100 ft or Ω /30.5 m)	Maximum Recommended Current (A)
14	0.2525	15
12	0.1588	20
10	0.9989	30
8	0.6282	55
6	0.3951	75
4	0.2485	95
3	0.1970	110
2	0.1563	130
1	0.1239	150
0	0.0983	170
00	0.0779	195
000	0.0618	225
0000	0.0490	260

Voltage drop due to line resistance:

$$\underbrace{\text{Voltage Drop}}_{\left(\frac{V}{100\,\text{ft}(30.5\text{m})}\right)} = \underbrace{\text{Current}}_{(A)} \times \underbrace{\text{Wire Resistance}}_{\left(\frac{\Omega}{100\,\text{ft}(30.5\text{m})}\right)}$$

Water Heating Systems

General design configurations and guidelines. Collector performance information is available on page 46, system evaluation and sizing can be estimated using the f-chart method, beginning page 66.

Integrated collector and storage or batch systems.

Storage tank integrated with collector or storage tank itself is the solar absorber. Circulation is passive through natural convection.

👍 Simple, no moving parts, long lifetime, little maintenance.

👎 Small systems only, limited freeze protection.

Natural circulation (thermosyphon loop) system.

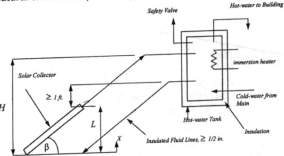

Circulation is caused by the difference in density, ρ, between the hot water in the collector and cooler water exiting the storage tank. To estimate circulation rate, compute the flow pressure drop, ΔP_{flow}:

$$\Delta P_{flow} = \rho_{storage} gH - \left[\rho_{collector\ ave} gL + \rho_{coll\ out} g(H-L) \right]$$

Flow velocity, V, can then be determined by knowing K_{loop}, the loop sum of the component velocity loss factors:

$$V = \sqrt{\frac{\Delta P_{flow}}{\rho_{loop\ ave} K_{loop}}}$$

Natural circulation (thermosyphon loop) system.

👍 Simple, moderate sizes, long lifetime in areas with little chance of freezing.

👎 Tank must be mounted above collectors, freeze protection difficult.

Forced circulation open loop system.

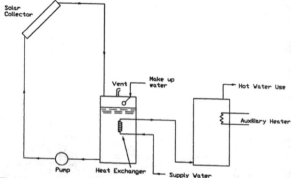

Fluid is actively pumped through the collector, and the reservoir is vented so pressure is maintained at atmospheric. Drainback operation is possible for freeze and stagnation protection.

👍 Simple, increased capacity compared to passive circulation.

👎 Freeze protection not as reliable as closed loop drainback, pump must supply entire head from storage tank to collector.

Forced circulation closed loop pressurized system.
Fluid loop is not vented so pumping power is limited to the flow resistance of the piping. Since fluid stays in the collector, glycol solutions are used for freeze protection.

👍 Good freeze protection in cold climates, reduced pumping power, can be used when drainback is not possible.

👎 Complex system, fluid expansion must be accommodated, no stagnation protection.

Forced circulation closed loop pressurized system.

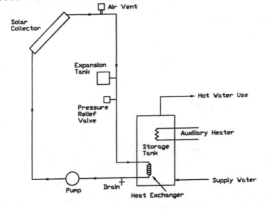

Forced circulation closed loop drainback system.

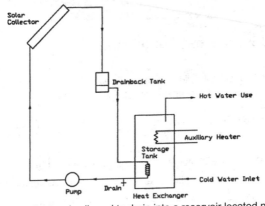

Fluid in collectors is allowed to drain into a reservoir located near the collectors but in a non-freezing location.

👍 Reliable freeze and stagnation protection, pure water working fluid can be used, pumping head reduced with elevated drainback tank.

👎 Piping must have sufficient slope for drainback.

Hot Water Load Data

Energy requirement for service water heating:

$$q_{hw} = \rho_w Q c_{pw} \left(T_d - T_s \right)$$

Where:

c_{pw}	specific heat of water, 4.18 kJ/kg-K
q_{hw}	water heating energy requirement
Q	volumetric water flow rate
T_d	water delivery temperature
T_s	water supply temperature
ρ_w	water density, 997 kg/m^3

Guidelines for service hot water demand rates

	Demand per person	
Usage type	liters/day	gal/day
Retail store	2.8	0.75
Elementary school	5.7	1.5
Multifamily residence	76.0	20.0
Single-family residence	76.0	20.0
Office building	11.0	3.0

Building Heat Load Data

Given the overall building loss coefficient, $\overline{UA}$ (W/°C), the degree-day method can be used to estimate the heat demand,

Q_n, for day n: $Q_n = \overline{UA} \left(T_{nl} - \overline{T}_a \right)_n$ with: $T_{nl} = T_i - \dfrac{q_i}{\overline{UA}}$

Where T_i is the interior temperature, q_i is the interior heat generation, and $\overline{T}_a$ (daily average temp.) can be estimated from location-specific maximum and minimum temperature data:

$$\overline{T}_a = \frac{T_{a,max} + T_{a,min}}{2}$$

Note: For a thorough treatment of heating loads, the reader is referred to the ASHRAE Handbook, Fundamentals, [1].

Heating System Evaluation and Sizing

Procedure for estimating the performance and/or size of standard solar heating applications using f-chart. The f-chart method computes the solar-supplied fraction, f_s, of thermal energy for liquid and air based heating systems.

Note: The procedures outlined here are for estimation purposes only, detailed calculations of the system thermal performance and solar collection are needed to ensure satisfactory system performance.

The f-chart method assumes standard system configurations, Figs. f-1 and f-2, and applies only to these systems, with limited variations. For example, the collector-to-storage heat exchanger in the liquid based system (Fig. f-1) may be eliminated ($F_{hx} =$); or for the air-based system (Fig. f-2) the two-tank domestic water heater may be reconfigured as a one tank system. Furthermore, f-chart is applicable to solar heating systems where the minimum temperature for energy delivery is approximately 20°C.

System parameter ranges used to compile the f-chart results:

Collector transmissivity-absorbtivity, $(\tau\alpha)_n$	0.6-0.9
Collector heat removal factor-area, $F_R A_c$	5-120 m^2
Collector heat loss coefficient, U_c	2.1-8.3 W/m^2-°C
Collector tilt angle, β	30°-90°
Overall building loss coefficient, $\overline{UA}$	83-667 W/°C

Begin with:

Monthly heating load; for information regarding water heating and building heating loads refer to page 65.

Monthly solar radiation totals for site-specific collector-plane; can be obtained from internet sources, page 83, or for simple geometries, computed with the procedure beginning on page 32.

Collector performance parameters $F_R \tau_s \alpha_s$ **and** $F_R U_c$;

can be obtained from the manufacturer or representative values for non-concentrating collectors are given on page 46.

Solar system design parameters; this includes the collector area, working fluid, fluid flow rate per unit area of collector, storage capacity, and heat exchanger performance. The standard configurations assumed with this method are shown in Figs. f-1 and f-2 for water and air based systems respectively.

Procedure begins on page 68.

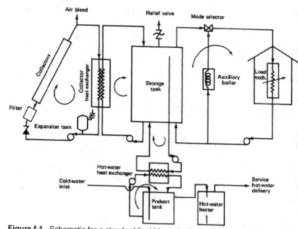

Figure f-1. Schematic for a standard liquid-based solar heating system. Note: certain deviations from this configuration can be handled by the f-chart method.

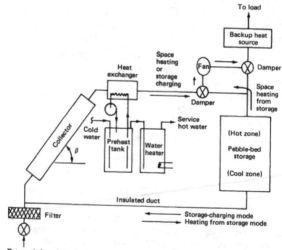

Figure f-2. Schematic for a standard air-based solar heating system. Note: certain deviations from this configuration can be handled by the f-chart method.

Procedure:

1. Compute the loss parameter, P_L:

$$P_L = \frac{A_c F_{hx} F_R U_c \Delta t (T_R - \overline{T}_a)}{L}$$

Where A_c is the net collector area (m^2), F_{hx} is the collector loop heat-exchanger factor (= 1 if no heat exchanger), Δt is the number of seconds per month, T_R is a reference temperature of 100° C, $\overline{T}_a$ is the monthly average ambient temperature (°C), and L is the total monthly heating load (J/month).

2. Compute the solar parameter, P_S:

$$P_s = \frac{A_c \overline{I}_c F F_{hx} F_R (\overline{\tau\alpha})_n}{L} \left(\frac{F_R (\overline{\tau\alpha})}{F_R (\tau\alpha)_n} \right)$$

Where $\overline{I}_c$ is the total monthly collector-plane insolation (J/m^2-month) and

$$\left(\frac{F_R (\overline{\tau\alpha})}{F_R (\tau\alpha)_n} \right) = 0.95 \text{ for collectors tilted within +/- 20° of the local latitude.}$$

3. Compute modified parameters based on deviations from standard systems. See Table f-1 for liquid based systems, Table f-2 for air based systems and/or the modification for water heating-only (below).

Water-heating-only loss parameter modification:

$$P_L = \frac{A_c F_{hx} F_R U_c \Delta t (11.6 + 1.18T_{w,o} + 3.86T_{w,i} - 2.32\overline{T}_a)}{L}$$

4. Compute the solar supplied fraction, f_s, of the monthly heating load. Read f_s from the appropriate figure, (Fig. f-3 liquid/Fig. f-4 air) or use the associated expressions (below):

Liquid based systems:

$$f_s = 1.029P_s - 0.065P_L - 0.245P_s^2 + 0.0018P_L^2 + 0.0215P_s^3$$

Air based systems:

$$f_s = 1.040P_s - 0.065P_L - 0.159P_s^2 + 0.00187P_L^2 - 0.0095P_s^3$$

For the conditions:

$$0 \leq P_s \leq 3.0; \quad 0 \leq P_L \leq 18.0; \quad 0 \leq f_s \leq 1.0 \quad \text{and} \quad \begin{array}{l} P_s > P_L/12 \text{ Liquid} \\ P_s > 0.07P_L \text{ Air} \end{array}$$

Table f-1. Liquid based system nominal values and modifying groups.

Parameter	Nominal value	Modified parameter[b]	
Flow rate[c] $\dfrac{(\dot{m}c_p)_c}{A_c}$	0.0128 liters H_2O equivalent/ sec $\cdot$ m$_c^2$	$P_L = P_{L,nom}\dfrac{F_{hx}F_R}{(F_{hx}F_R)_{nom}}$	(5.23)
		$P_s = P_{s,nom}\dfrac{F_{hx}F_R}{(F_{hx}F_R)_{nom}}$	(5.24)
Storage volume (water) $V_s = \left(\dfrac{M}{\rho A_c}\right)_s$	75 liters H_2O/m$_c^2$	$P_L = P_{L,nom}\left(\dfrac{V_s}{75}\right)^{-0.25}$	(5.25)
Load heat exchanger[d] $\dfrac{\epsilon_L(\dot{m}c_p)_{air}}{Q_L}$	2.0	$P_s = P_{s,nom}\left\{0.393 + 0.651\exp\left[-0.139\dfrac{Q_L}{\epsilon_L(\dot{m}c_p)_{min}}\right]\right\}$	(5.26)

[c]Table prepared from data and equations presented in [2,8]
[b]Multiply basic definition of P_s and P_L in points 1 and 2 by factor for nonnominal group values; $(F_{hx}F_R)_{nom}$ refers to values of $F_{hx}F_R$ at collector rating or test conditions.
[c]In liquid systems the correction for flow rate is small and can usually be ignored if variation is no more than 50 percent below the nominal value.
[d]$(\dot{m}c_p)_{min}$ is the minimum fluid capacitance rate, usually that of air for the load heat exchanger; Q_L is the heat load per unit temperature difference between inside and outside of the building.

Table f-2. Air based system nominal values and modifying factors.

Parameter	Nominal value[b]	Loss parameter multiplier
Storage capacity V_s	0.25 m³/m$_c^2$	$\left(\dfrac{0.25}{V_s}\right)^{0.3}$
Fluid volumetric flow rate Q_c	10.1 liters/sec $\cdot$ m$_c^2$	$\left(\dfrac{Q_c}{10.1}\right)^{0.28}$

[a]Adapted from [2,8]
[b]Based on net collector area; fluid volume at standard atmosphere conditions.

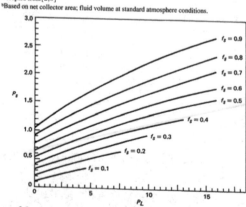

Figure f-3. f-chart for liquid based solar heating systems, [8].

69

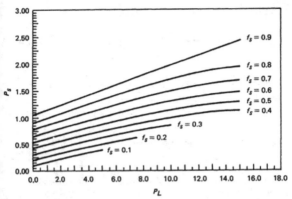

Figure f-4. f-chart for air based solar heating systems, [8].

Daylighting

General design guidelines, from [7]

- Orient building to maximize daylighting. Long axis running east-west preferred.
- Maximize south glazing, minimize east and west facing glass. A south-facing aperture is the only orientation that, on an annual basis, balances typical thermal needs and lighting requirements with available radiation.
- Optimally size overhangs on south-facing glazing to harvest daylighting, reduce summer heat gain, and permit the passive collection of solar thermal in winter.
- Select the right glazing. Where windows are used specifically for daylighting, clear glass has an advantage over glazing with a low-E coating due to the coating's typical 10% to 30% reduction in visible light transmission.
- Eliminate direct beam radiation. Use baffles to block direct beam radiation, diffuse light, and reduce glare.
- Account for shading from adjacent buildings and trees and consider the reflectance from adjacent surfaces.
- Use light colored roofing in front of monitors and select light colors for interior finishes to reflect in additional light and enhance distribution throughout the room

70

Daylighting Comparison

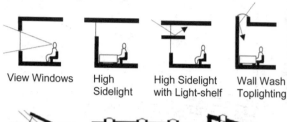

View Windows High Sidelight High Sidelight with Light-shelf Wall Wash Toplighting

 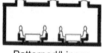

Central Toplighting Patterned/Linear Toplighting Tubular Skylights

Selection Criteria for Daylighting Strategies.

Design Criteria	View Windows (DL1)	High Sidelight w/ Light Shelf (DL2 & DL3)	Wall Wash Toplighting (DL4)	Central & Patterned Toplighting (DL5 & DL6)	Linear Toplighting (DL7)	Tubular Skylights (DL8)
Uniform Light Distribution	◐◐	●/○	●	●●	◐	◐
Low Glare	○	●	●	●	●	●/○
Reduced Energy Costs	○	●	●	●●	●	●
Cost Effectiveness	●	●	●	●	●	●●
Safety/Security Concerns	○	●/○	●	●	●	●●
Low Maintenance	○	●	●	●	●	●

●● Extremely good application ● Good application ○ Poor application ◐◐ Extremely poor application ◐ Depends on space layout and number and distribution of daylight apertures ●/○ Mixed benefits

Reprinted from [14

Note: The configuration of the view window is with no controls, daylighting performance can be improved with appropriate measures, overhangs, shades, etc. Additionally, the benefit of providing a direct visual connection to the outdoors is not considered in this comparison.

Daylighting Design

Lumen method for estimating workplane illuminance level with sidelighting and skylighting.

Sidelighting with vertical windows:

The lumen method computes work plane illuminance levels for five reference points for the standard geometry shown in the figure d-1, next page.

Sidelighting with vertical windows:

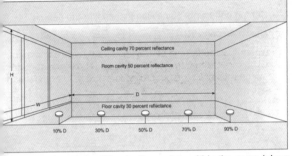

Figure d-1. Location of illumination points within the room (along the centerline of window) determined by lumen method of sidelighting, [6].

. **Determine the total sky illuminance entering the window,** E_{sw}. Compute the solar altitude and azimuth angle for the desired latitude, date, and time of day, page 6. Compute the sun-window normal azimuth angle difference using: $a_{sw} = |a_s - a_w|$ Using the figures for vertical surfaces on page 74, determine the direct sun illuminance, $E_{v,sun}$, and the direct sky illuminance, $E_{v,sky}$, for the appropriate sky conditions. Then: $E_{sw} = E_{v,sun} + E_{v,sky}$.

. **Determine the reflected ground illuminance entering the window,** E_{gw}. Read values of direct sun and sky illuminance for a horizontal surface, $E_{h,sun}$ and $E_{h,sky}$, page 74. For uniformly reflective ground surfaces extending from the window outward to the horizon, the illuminance on the window from ground reflection, E_{gw}, can be determined with: $E_{gw} = \rho \left(E_{h,sun} + E_{h,sky} \right) / 2$ Where typical reflectivity values, ρ, come from the table on page 36.

. Determine the net window transmittance, τ.

$\tau = TR_a T_c LLF$ where T is the glazing transmittance, table d-1, R_A is the ratio of net to gross window areas and T_C is the transmittance of any light-controlling devices. Light loss factor values, LLF, can be taken from table d-2.

72

Sidelighting with vertical windows:

4. Compute the work plane illuminances for the geometry shown on page 72. $E_{TWP} = \tau \left(E_{sw} CU_{sky} + E_{gw} CU_g \right)$

where the coefficients of utilization for the sky component, CU_{sky}, are taken from Table d-3, page 75-76, a-e depending on the ratio of E_{vsky}/E_{hsky}. CU_g values are read from table d-3, f.

Table d-1. Glass transmittances

Glass	Thickness (in.)	τ
Clear	$\frac{1}{8}$	.89
Clear	$\frac{3}{16}$	.88
Clear	$\frac{1}{4}$	.87
Clear	$\frac{5}{16}$	.86
Grey	$\frac{1}{8}$	.61
Grey	$\frac{3}{16}$	.51
Grey	$\frac{1}{4}$	.44
Grey	$\frac{5}{16}$	.35
Bronze	$\frac{1}{8}$	.68
Bronze	$\frac{3}{16}$	.59
Bronze	$\frac{1}{4}$	.52
Bronze	$\frac{5}{16}$	.44
Thermopane	$\frac{1}{8}$	.80
Thermopane	$\frac{3}{16}$	.79
Thermopane	$\frac{1}{4}$	.77

Murdoch [11]

Table d-2. Light loss factors, [6].

Locations	Light Loss Factor Glazing Position		
	Vertical	Sloped	Horizontal
Clean Areas	0.9	0.8	0.7
Industrial Areas	0.8	0.7	0.6
Very Dirty Areas	0.7	0.6	0.5

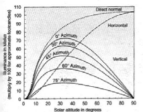

Figure d-2. a. Clear sky, direct sun illuminance, vertical surface, [6]

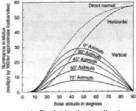

b. Partly cloudy sky, direct sun illuminance, vertical surface, [6]

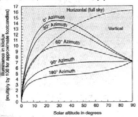

c. Clear sky, sky illuminance, vertical surface, [6]

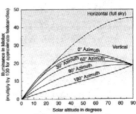

d. Partly cloudy sky, sky illuminance, vertical surface, [6]

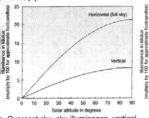

e. Overcast sky, sky illuminance, vertical surface, [6]

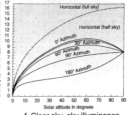

f. Clear sky, sky illuminance, horizontal half sky, [6]

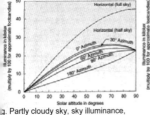

g. Partly cloudy sky, sky illuminance, horizontal half sky, [6]

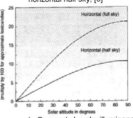

h. Overcast sky, sky illuminance, horizontal half sky, [6]

74

Sidelighting with vertical windows:

Table d-3. Coefficients of utilization for lumen method of sidelighting (window without blinds), [6].

Room Depth/ Window Height	Percent D*	.5	1	2	3	4	6	8	Infinite
		Window Width/Window Height							
1	10	.824	.864	.870	.873	.875	.879	.880	.883
	30	.547	.711	.777	.789	.793	.798	.799	.801
	50	.355	.526	.635	.659	.666	.669	.670	.672
	70	.243	.386	.505	.538	.548	.544	.545	.547
	90	.185	.304	.418	.451	.464	.444	.446	.447
2	10	.667	.781	.809	.812	.813	.815	.816	.824
	30	.269	.416	.519	.544	.551	.556	.557	.563
	50	.122	.204	.287	.319	.331	.339	.341	.345
	70	.068	.116	.173	.201	.214	.223	.226	.229
	90	.050	.084	.127	.151	.164	.167	.171	.172
3	10	.522	.661	.739	.746	.747	.749	.747	.766
	30	.139	.232	.320	.350	.360	.366	.364	.373
	50	.063	.092	.139	.163	.174	.183	.182	.187
	70	.031	.053	.081	.097	.106	.116	.116	.119
	90	.025	.041	.061	.074	.082	.089	.090	.092
4	10	.405	.576	.658	.670	.673	.675	.674	.707
	30	.075	.134	.197	.224	.235	.243	.243	.255
	50	.028	.050	.078	.094	.104	.112	.114	.119
	70	.018	.031	.048	.059	.065	.073	.074	.078
	90	.016	.026	.040	.048	.053	.059	.061	.064
6	10	.242	.392	.494	.516	.521	.524	.523	.588
	30	.027	.054	.086	.102	.111	.119	.120	.135
	50	.011	.023	.036	.044	.049	.055	.056	.063
	70	.009	.018	.027	.032	.035	.040	.041	.046
	90	.008	.016	.023	.028	.031	.034	.035	.040
8	10	.147	.257	.352	.380	.387	.391	.392	.482
	30	.012	.026	.043	.054	.060	.067	.070	.086
	50	.006	.013	.021	.026	.029	.033	.035	.043
	70	.005	.011	.017	.021	.023	.026	.027	.034
	90	.004	.010	.015	.019	.021	.023	.025	.030
10	10	.092	.168	.248	.275	.284	.290	.291	.395
	30	.006	.014	.026	.032	.036	.041	.044	.059
	50	.003	.008	.014	.017	.019	.022	.024	.032
	70	.003	.007	.012	.014	.016	.018	.019	.026
	90	.003	.006	.011	.013	.015	.016	.017	.024

*Percent D is the relative distance from the window to the opposite wall.

a. CUsky for (Ev/Eh)sky = 0.75

Room Depth/ Window Height	Percent D*	.5	1	2	3	4	6	6	Infinite
		Window Width/Window Height							
1	10	.671	.704	.711	.715	.717	.726	.726	.728
	30	.458	.595	.654	.668	.672	.682	.683	.685
	50	.313	.462	.563	.589	.598	.607	.608	.610
	70	.227	.362	.478	.515	.527	.530	.532	.534
	90	.186	.306	.424	.465	.481	.468	.471	.472
2	10	.545	.636	.658	.660	.661	.665	.666	.672
	30	.239	.367	.459	.484	.491	.499	.501	.506
	50	.121	.203	.286	.320	.335	.348	.351	.355
	70	.074	.128	.192	.226	.243	.259	.264	.267
	90	.058	.101	.156	.188	.207	.215	.221	.223
3	10	.431	.561	.607	.613	.614	.616	.615	.631
	30	.133	.223	.306	.337	.348	.357	.357	.366
	50	.058	.103	.155	.183	.197	.211	.213	.218
	70	.037	.064	.098	.119	.132	.147	.150	.154
	90	.030	.051	.079	.098	.110	.122	.126	.129
4	10	.339	.482	.549	.560	.563	.566	.565	.593
	30	.078	.139	.204	.234	.247	.258	.260	.272
	50	.033	.060	.094	.114	.126	.139	.143	.150
	70	.022	.039	.061	.074	.083	.095	.099	.104
	90	.019	.032	.050	.061	.070	.080	.084	.089
6	10	.211	.343	.433	.453	.458	.461	.461	.518
	30	.033	.065	.103	.123	.135	.145	.148	.167
	50	.015	.029	.047	.057	.064	.073	.077	.086
	70	.011	.021	.033	.040	.045	.051	.054	.060
	90	.010	.019	.028	.034	.038	.044	.046	.052
8	10	.135	.238	.326	.353	.362	.366	.367	.452
	30	.016	.034	.058	.072	.080	.090	.094	.116
	50	.008	.017	.027	.034	.039	.045	.048	.059
	70	.006	.013	.021	.026	.028	.032	.035	.043
	90	.005	.012	.019	.023	.025	.029	.031	.038
10	10	.090	.165	.244	.272	.283	.290	.291	.395
	30	.009	.020	.036	.045	.052	.060	.064	.087
	50	.005	.010	.019	.023	.026	.030	.033	.044
	70	.004	.009	.015	.018	.020	.023	.025	.033
	90	.003	.008	.014	.016	.018	.020	.022	.030

*Percent D is the relative distance from the window to the opposite wall.

b. CUsky for (Ev/Eh)sky = 1.00

Room Depth/ Window Height	Percent D*	.5	1	2	3	4	6	8	Infinite
		Window Width/Window Height							
1	10	.578	.607	.614	.619	.621	.633	.634	.635
	30	.406	.525	.580	.594	.599	.612	.614	.615
	50	.287	.423	.519	.547	.556	.569	.571	.573
	70	.218	.347	.461	.501	.515	.522	.525	.526
	90	.186	.307	.428	.473	.491	.483	.486	.487
2	10	.472	.549	.566	.569	.570	.574	.575	.581
	30	.221	.337	.422	.447	.456	.465	.467	.472
	50	.120	.202	.285	.321	.337	.353	.357	.361
	70	.078	.136	.204	.242	.261	.281	.287	.290
	90	.064	.112	.174	.211	.233	.244	.251	.253
3	10	.377	.488	.527	.533	.534	.536	.536	.549
	30	.130	.217	.298	.329	.341	.352	.353	.362
	50	.062	.110	.165	.195	.211	.228	.231	.237
	70	.040	.070	.109	.132	.147	.166	.171	.175
	90	.033	.057	.090	.112	.127	.142	.148	.152
4	10	.300	.424	.484	.494	.497	.499	.499	.524
	30	.080	.143	.209	.240	.255	.267	.269	.283
	50	.036	.066	.104	.126	.140	.156	.160	.168
	70	.024	.043	.068	.083	.094	.109	.115	.120
	90	.021	.036	.056	.070	.080	.092	.099	.103
6	10	.193	.314	.395	.415	.420	.423	.423	.476
	30	.036	.071	.113	.136	.149	.161	.165	.186
	50	.017	.033	.053	.065	.074	.084	.089	.100
	70	.012	.024	.037	.045	.050	.058	.061	.069
	90	.011	.021	.031	.038	.043	.049	.053	.060
8	10	.128	.226	.310	.337	.346	.351	.352	.433
	30	.019	.039	.066	.082	.092	.104	.109	.134
	50	.009	.019	.031	.040	.045	.052	.056	.069
	70	.007	.015	.023	.029	.032	.037	.040	.049
	90	.006	.013	.021	.025	.028	.032	.035	.043
10	10	.088	.164	.241	.270	.282	.290	.291	.396
	30	.011	.024	.043	.054	.062	.071	.076	.103
	50	.005	.012	.022	.026	.030	.035	.038	.052
	70	.004	.010	.017	.020	.023	.026	.028	.038
	90	.004	.009	.016	.018	.020	.023	.025	.034

*Percent D is the relative distance from the window to the opposite wall.

c. CUsky for (Ev/Eh)sky = 1.25

Table d-3. Coefficients of utilization for lumen method of Sidelighting (window without blinds), [6].

Room Infinite Depth/Window Height	Percent D*	Window Width/Window Height							
		.5	1	2	3	4	6	8	Infinite
1	10	105	137	177	197	207	208	210	211
	30	116	157	205	228	235	241	243	244
	50	104	165	217	243	253	283	285	286
	70	101	182	217	243	253	283	285	286
	90	091	148	199	230	239	290	292	293
2	10	085	124	180	178	186	189	189	191
	30	082	132	179	201	212	219	222	225
	50	062	113	165	189	202	214	218	220
	70	051	093	141	165	178	179	183	185
	90	045	079	118	140	153	179	183	185

f. CUg (ground reflected component)

e. CUsky for (Ev/Eh)sky = 1.75

d. CUsky for (Ev/Eh)sky = 1.50

Lumen method for skylighting:

1. **Determine the total sky illuminance entering the skylight.** Compute the solar altitude angle for the desired latitude, date, and time of day, page 6. Determine the total horizontal skylight illuminance, E_H, (direct sun plus sky illuminance) using the appropriate horizontal surface data in Figure d-2, page 74.

2. **Determine the net skylight transmittance, T_n.** The flat plate transmittances, T_F, for several plastic materials are provided in Table d-3 below. For domed skylights, the effective dome transmittance, T_D, can be computed using:

   $$T_D = 1.25 T_F \left(1.18 - 0.416 T_F \right)$$ For double-domed

   skylights, the individual transmittances can be combined

   using: $$T_D = \frac{T_{D1} T_{D2}}{T_{D1} T_{D2} - T_{D1} T_{D2}}$$ Also determine the light loss

   factor, LLF from Table d-2, page 73. Compute the well

   cavity ratio for the skylight well with: $$WCR = \frac{5h(w+l)}{wl}$$

   Using Figure d-3, page 78, determine the skylight well efficiency, N_w. Compute the net skylight transmittance, T_n, using: $T_n = T_D N_w R_A T_C LLF$ where R_A is the ratio of net to gross skylight areas and T_C is the transmittance of any light-controlling devices.

Table d-3. Flat-plate plastic material transmittance for skylights. Source: Murdoch[11]

Type	Thickness (in.)	Transmittance
Transparent	$\frac{1}{8} - \frac{3}{16}$	.92
Dense translucent	$\frac{1}{8}$	.32
Dense translucent	$\frac{3}{16}$	.24
Medium translucent	$\frac{1}{8}$	.56
Medium translucent	$\frac{3}{16}$	.52
Light translucent	$\frac{1}{8}$	.72
Light translucent	$\frac{3}{16}$	.68

Lumen method for skylighting:

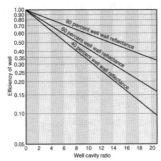

Figure d-3. Efficiency of well, N_w, versus well cavity ratio, [6].

3. **Compute the room coefficient of utilization, CU.** For office and warehouse interiors CU can be estimated with:

$$CU = \frac{1}{1 + A\left(RCR\right)^B} \quad \text{for } RCR < 8 \quad \text{where}$$

A=0.0288, B=1.560 for offices (typical reflectance: ceiling 0.75, wall 0.5, floor 0.3) and A=0.0995, B=1.087 for warehouses (typical reflectance: ceiling 0.5, wall 0.3, floor 0.2). The room cavity ratio is given by

$$RCR = \frac{5h_c\left(l + w\right)}{lw} \quad \text{with } h_c \text{ being the ceiling height}$$

above the work plane and l and w being the room length and width, respectively.

4. **Compute the illuminance at the work plane, E_{TWP}:**

$$E_{TWP} = E_H T_n \left(CU\right)\left(\frac{A_T}{A_{WP}}\right) \quad \text{:where } A_T \text{ is the total gross}$$

area of the skylights (number of skylights times the skylight gross area), and A_{WP} is the work plane area (typically room length times width). Note that it is possible to fix the E_{TWP} at some desired value and determine the skylight area required, however, due to their dependence on A_T, the factors N_w and R_A (step 2) should be recalculated.

78

Solar Dryer Configurations

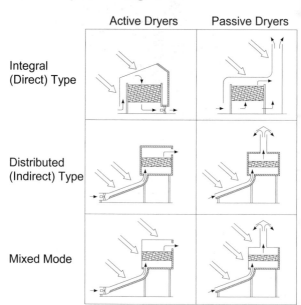

Adapted from [12]

Safe storage moisture for aerated good quality grain, [15].

Grain	Maximum safe moisture content
Shelled corn and sorghum	
To be sold as #2 grain or equiv. by spring	15.5%
To be stored up to 1 year	14%
To be stored more than 1 year	13%
Soybeans	
To be sold by spring	14%
To be stored up to 1 year	12%
Wheat	13%
Small grain (oats, barley, etc.)	13%
Sunflowers	9%

Life Cycle Costing

Life Cycle Cost (LCC) is an economic measure that reflects the benefits accrued by solar usage throughout the lifetime of a solar-powered system. This is opposed to a simple initial cost comparison, which does not include lifetime energy consumption and, therefore, typically favors non-renewable alternatives.

LCC is the sum of the present worth (pw) values of all of the expenses associated with the system over its expected lifetime:

$$LCC = C + M_{pw} + E_{pw} + R_{pw} - S_{pw}$$

Where:

C   capital cost of a project which includes the initial capital expense for equipment, the system design, engineering, and installation. This cost is always considered as a single payment occurring in the initial year of the project, regardless of how the project is financed.

M   maintenance is the sum of all yearly scheduled operation and maintenance (O&M) costs. Fuel or equipment replacement costs are not included.

E   energy cost is the sum of the yearly fuel or electricity cost for the system.

R   replacement cost is the sum of all repair and equipment replacement cost anticipated over the life of the system.

S   salvage value of a system is its net worth in the final year of the life-cycle period. It is common practice to assign a salvage value of 20% of original cost for mechanical equipment that can be moved.

Computation of the present worth of future expenditures.

1. The single present worth (P) of a future sum of money (F) in a given year (N) at a given investment rate (D) and inflation rate (i) is:

$$P = FX^N \quad \text{with} \quad X = \left(\frac{1+i}{1+D} \right)$$

2. The uniform present worth (P) of an annual sum (A) received over a period of years (N) at a given investment rate (D) and inflation rate (i) is:

$$P = A\left(1 - X^N\right)/\left(X^{-1} - 1\right)$$

Example: Compare the life cycle cost for a photovoltaic power supply with battery storage and a gasoline engine-generator alternative (systems specified below).

Given: Life cycle period: 20 years
        Investment rate: 7%
        General inflation: 3%
        Fuel inflation: 4%

Solution: Begin with system designs for each alternative, making sure that each system provides equivalent performance, for example: power output, reliability, lifetime, etc. Using a detailed cost estimate (sample below), the present worth of each component is determined. For future one-time expenditures, (e.g. battery replacement, generator rebuild) point 1 above can be used. For costs repeating on an annual basis (e.g. maintenance, generator fuel) point 2 is used. Note the use of a separate fuel inflation rate for fuel expenses. The present worth values are then summed (less salvage) to give the LCC.

PV System

Item	Initial Cost ($)	Present Worth ($)
1. Capital		
Array	2500	2500
Controller	300	300
Batteries	900	900
Installation	700	700
2. Maintenance		
Annual Inspection (per year)	75	1030
3. Energy		
None	-	-

continued

Item	Initial Cost ($)	Present Worth ($)
4. Replacement		
Battery bank @ yr. 5	900	744
Battery bank @ yr. 10	900	615
Battery bank @ yr. 15	900	509
5. Salvage		
20% of original equipment cost	740	(346)
LCC: (Items 1 + 2 + 3 + 4 - 5)		$6,952

Engine-generator System

Item	Initial Cost ($)	Present Worth ($)
1. Capital		
Generator	400	400
Installation	300	300
2. Maintenance		
Tune-up (per year)	150	2060
Annual Inspection (per year)	75	1030
3. Energy		
Annual fuel cost (4% fuel inflation rate)	375	5640
4. Replacement		
Gen. rebuild @ yr. 5	250	207
Gen. rebuild @ yr. 10	250	171
Gen. rebuild @ yr. 15	250	142
5. Salvage		
20% of original equipment cost	80	(38)
LCC: (Items 1 + 2 + 3 + 4 - 5)		$9,912

References

Internet sources of data

Solar Resource Data

World radiation data center (WRDC) online archive, Russian Federal Service for Hydrometeorology and Environmental Monitoring; 1964-1993 data http://wrdc-mgo.nrel.gov/ 1994-present data http://wrdc.mgo.rssi.ru/

Surface meteorology and solar energy, National Aeronautics and Space Administration, USA; http://eosweb.larc.nasa.gov/sse

Solar radiation resource information, National Renewable Energy Laboratory, USA; http://rredc.nrel.gov/solar

Climatic Data

World climatic data, World Weather Information Service; http://www.worldweather.org

U. S. climate data, National Oceanic and Atmospheric Administration, USA; http://www.noaa.gov/climate.html

Cited References

ASHRAE. 2001. *2001 ASHRAE Handbook, Fundamentals*. Atlanta: American Society of Heating, Refrigerating and Air-Conditioning Engineers, Inc.

Duffie, J. A. and Beckman, W. A. 1991. *Solar Engineering of Thermal Processes*. 2nd ed. New York: John Wiley & Sons.

Building Technologies Program, Lawrence Berkeley National Lab. 1997. *Tips for Daylighting with Windows*. Berkeley: Lawrence Berkeley National Lab.

Collares-Pereira and Rabl, A. 1979. Simple Procedure for Predicting Long Term Average Performance of Nonconcentrating and of Concentrating Solar Collectors. *Solar Energy*, Vol. 23, pg. 235-254.

Goswami, D. Y., Kreith, F., and Kreider, J. 2000. *Principles of .Solar Engineering*, 2nd ed. Philadelphia: Taylor & Francis.

IESNA. 2000. *The IESNA Lighting Handbook*. New York: Illumination Engineering Society of North America. Reprinted with permission from the IESNA Lighting Handbook, 9th Edition, courtesy of the Illuminating Engineering Society of North America.

Innovative Design, Inc. 2004. *Guide for Daylighting Schools*. Raleigh, NC: Innovative Design, Inc.

Klein, S. A. 1976. *A Design Procedure for Solar Heating Systems*, Ph.D. dissertation, Univ. of Wisconsin, Madison. For an approach similar to the f-chart for other solar-thermal systems operating above a minimum temperature above that for space-heating (~20°C), see Klein, S. A. and Beckman, W. A. 1977. A General Design Method for Closed Loop Solar Energy Systems. *Proc. 1977 ISES Meeting*.

Löf, G. O. G. and Tybout, R. A. 1972. Model for Optimizing Solar Heating Design. ASME Paper 72-WA/SOL-8.

10. Messenger, R. and Ventre, J. 2000. *Photovoltaic Systems Engineering*. Boca Raton, FL: CRC Press.

1. Murdoch, J. B. 1985. *Illumination Engineering: From Edison's Lamp to the Laser*. New York: Macmillan Publishing Co.

2. Norton, B. 1992. *Solar Energy Thermal Technology*. London: Springer-Verlag.

3. Post, H. N. and Risser, V. V. 1995. *Stand-Alone Photovoltaic Systems—A Handbook of Recommended Design Practices*. Albuquerque: Sandia National Lab. Report SAND-87-7023.

4. U.S. Department of Energy. 2002. *National Best Practices Manual for Building High Performance Schools*. Publication DOE/GO-102002-1610.

5. Midwest Plan Service. 1980. *Low Temperature and Solar Grain Drying Handbook*. Ames, IA: Iowa State University.

Units and Conversion Factors

Fundamental SI units

Quantity	Name of unit	Symbol
Length	Meter	m
Mass	Kilogram	kg
Time	Second	sec
Electric current	Ampere	A
Thermodynamic temperature	Kelvin	K
Luminous intensity	Candela	cd
Amount of a substance	Mole	mol

Derived SI units

Quantity	Name of unit	Symbol
Acceleration	Meters per second squared	m/sec^2
Area	Square meters	m^2
Density	Kilogram per cubic meter	kg/m^3
Dynamic viscosity	Newton-second per square meter	$N \cdot sec/m^2$
Force	Newton (= 1 kg $\cdot$ m/sec^2)	N
Frequency	Hertz	Hz
Kinematic viscosity	Square meter per second	m^2/sec
Plane angle	Radian	rad
Potential difference	Volt	V
Power	Watt (= 1 J/s)	W
Pressure	Pascal (= 1 N/m^2)	Pa
Radiant intensity	Watts per steradian	W/sr
Solid angle	Steradian	sr
Specific heat	Joules per kilogram-Kelvin	$J/kg \cdot K$
Thermal conductivity	Watts per meter-Kelvin	$W/m \cdot K$
Velocity	Meters per second	m/sec
Volume	Cubic meter	m^3
Work, energy, heat	Joule (= 1 N $\cdot$ m)	J

Fundamental constants

Quantity	Symbol	Value
Avogadro constant	N	$6.022169 \times 10^{26}\ kmol^{-1}$
Boltzmann constant	k	$1.380622 \times 10^{-23}\ J/K$
First radiation constant	$C_1 = 2\pi hc^2$	$3.741844 \times 10^{-16}\ W \cdot m^2$
Gas constant	R	$8.31434 \times 10^3\ J/kmol \cdot K$
Planck constant	h	$6.626196 \times 10^{-34}\ J \cdot sec$
Second radiation constant	$C_2 = hc/k$	$1.438833 \times 10^{-2}\ m \cdot K$
Speed of light in a vacuum	c	$2.997925 \times 10^8\ m/sec$
Stefan-Boltzmann constant	σ	$5.66961 \times 10^{-8}\ W/m^2 \cdot K^4$

Conversion Factors

Physical quantity	Symbol	Conversion factor
Area	A	$1\ \text{ft}^2 = 0.0929\ \text{m}^2$
		$1\ \text{acre} = 43{,}560\ \text{ft}^2 = 4047\ \text{m}^2$
		$1\ \text{hectare} = 10{,}000\ \text{m}^2$
		$1\ \text{square mile} = 640\ \text{acres}$
Density	ρ	$1\ \text{lb}_m/\text{ft}^3 = 16.018\ \text{kg/m}^3$
Heat, energy, or work	Q or W	$1\ \text{Btu} = 1055.1\ \text{J}$
		$1\ \text{kWh} = 3.6\ \text{MJ}$
		$1\ \text{Therm} = 105.506\ \text{MJ}$
		$1\ \text{cal} = 4.186\ \text{J}$
		$1\ \text{ft} \cdot \text{lb}_f = 1.3558\ \text{J}$
Force	F	$1\ \text{lb}_f = 4.448\ \text{N}$
Heat flow rate, Refrigeration	q	$1\ \text{Btu/hr} = 0.2931\ \text{W}$
		$1\ \text{ton (refrigeration)} = 3.517\ \text{kW}$
		$1\ \text{Btu/sec} = 1055.1\ \text{W}$
Heat flux	q/A	$1\ \text{Btu/hr} \cdot \text{ft}^2 = 3.1525\ \text{W/m}^2$
Heat-transfer coefficient	h	$1\ \text{Btu/hr} \cdot \text{ft}^2 \cdot \text{F} = 5.678\ \text{W/m}^2 \cdot \text{K}$
Length	L	$1\ \text{ft} = 0.3048\ \text{m}$
		$1\ \text{in} = 2.54\ \text{cm}$
		$1\ \text{mi} = 1.6093\ \text{km}$
Mass	m	$1\ \text{lb}_m = 0.4536\ \text{kg}$
		$1\ \text{ton} = 2240\ \text{lbm}$
		$1\ \text{tonne (metric)} = 1000\ \text{kg}$
Mass flow rate	$\dot{m}$	$1\ \text{lb}_m/\text{hr} = 0.000126\ \text{kg/sec}$
Power	$\dot{W}$	$1\ \text{hp} = 745.7\ \text{W}$
		$1\ \text{kW} = 3415\ \text{Btu/hr}$
		$1\ \text{ft} \cdot \text{lb}/\text{sec} = 1.3558\ \text{W}$
		$1\ \text{Btu/hr} = 0.293\ \text{W}$
Pressure	p	$1\ \text{lb}/\text{in}^2\ (\text{psi}) = 6894.8\ \text{Pa (N/m}^2)$
		$1\ \text{in. Hg} = 3{,}386\ \text{Pa}$
		$1\ \text{atm} = 101{,}302\ \text{Pa (N/m}^2) = 14.696\ \text{psi}$
Radiation	l	$1\ \text{langley} = 41{,}860\ \text{J/m}^2$
		$1\ \text{langley/min} = 697.4\ \text{W/m}^2$
Specific heat capacity	c	$1\ \text{Btu/lb}_m \cdot {}^\circ\text{F} = 4187\ \text{J/kg} \cdot \text{K}$
Internal energy or enthalpy	e or h	$1\ \text{Btu/lb}_m = 2326.0\ \text{J/kg}$
		$1\ \text{cal/g} = 4184\ \text{J/kg}$
Temperature	T	$T(^\circ\text{R}) = (9/5)T(\text{K})$
		$T(^\circ\text{F}) = [T(^\circ\text{C})](9/5) + 32$
		$T(^\circ\text{F}) = [T(\text{K}) - 273.15](9/5) + 32$
Thermal conductivity	k	$1\ \text{Btu/hr} \cdot \text{ft} \cdot {}^\circ\text{F} = 1.731\ \text{W/m} \cdot \text{K}$
Thermal resistance	R_{th}	$1\ \text{hr} \cdot {}^\circ\text{F/Btu} = 1.8958\ \text{K/W}$
Velocity	V	$1\ \text{ft/sec} = 0.3048\ \text{m/sec}$
		$1\ \text{mi/hr} = 0.44703\ \text{m/sec}$
Viscosity, dynamic	μ	$1\ \text{lb}_m/\text{ft} \cdot \text{sec} = 1.488\ \text{N} \cdot \text{sec/m}^2$
		$1\ \text{cP} = 0.00100\ \text{N} \cdot \text{sec/m}^2$
Viscosity, kinematic	v	$1\ \text{ft}^2/\text{sec} = 0.09029\ \text{m}^2/\text{sec}$
		$1\ \text{ft}^2/\text{hr} = 2.581 \times 10^{-5}\ \text{m}^2/\text{sec}$
Volume	V	$1\ \text{ft}^3 = 0.02832\ \text{m}^3 = 28.32\ \text{liters}$
		$1\ \text{barrel} = 42\ \text{gal (U.S.)}$
		$1\ \text{gal (U.S. liq.)} = 3.785\ \text{liters}$
		$1\ \text{gal (U.K.)} = 4.546\ \text{liters}$
Volumetric flow rate	$\dot{Q}$	$1\ \text{ft}^3/\text{min (cfm)} = 0.000472\ \text{m}^3/\text{sec}$
		$1\ \text{gal/min (GPM)} = 0.0631\ \text{l/sec}$

The History of ISES

1954 ISES has its origin in Phoenix, Arizona, USA. A group of industrial, financial and agricultural leaders establishes the "Association for Applied Solar Energy" (AFASE) as a non-profit organization.

1955 The first two important meetings are held in Tucson and Phoenix, USA, which attract more than 1000 scientists, engineers and government officials from 36 different countries.

1956 The association establishes its first scientific publication "the sun at work".

1957 The first issue of "The Journal of Solar Energy, Science and Engineering" is published.

1963 Dedicated solar scientists decide that radical changes are required for the operation and goals of the society. Through the reorganization within the framework of its original concept the name is changed to "The Solar Energy Society". Accreditation of the society at the United Nations Economic & Social Council (ECOSOC).

1964 The name of the journal is changed to "Solar Energy – The Journal of Solar Energy Science and Technology".
Prof. Farrington Daniels is elected President and in his honor the 'Farrington Daniels Award' is established in 1975. The award dignifies outstanding intellectual leadership in the field of renewable energy.

1970 Relocation of the international headquarters office to Melbourne, Australia. First international conference outside the USA is held in Melbourne, Australia.

1971 The name of the society is changed to "International Solar Energy Society".

1976 The first issue of the ISES magazine "SunWorld" is published.

1979 ISES celebrates its Silver Jubilee at the Solar World Congress in Atlanta, USA.

1980 The 'Achievement through Action Award' in memory of Christopher A. Weeks is set up. A substantial cash prize honors contributions for practical use or new concepts.

1989 ISES introduces the "Section Sponsorship Programme" in which individuals and sections have the opportunity to sponsor sections from developing countries.

1992 Accepted by the United Nations as a non-governmental organization (NGO) in consultative status, ISES actively participates in the United Nations Conference on Environment and Development (UNCED), held in Rio de Janeiro, Brazil.

1994 ISES coordinates the second meeting of the United Nations Commission on Sustainable Development (CSD) in New York to discuss issues of sustainable development and the interlinkages to renewable energy technologies.

1995 The Headquarters office moves from Melbourne, Australia to Freiburg, Germany. The Headquarters becomes a focal point for international projects.

2000 Launch of new official ISES magazine "Refocus" (Renewable Energy Focus).

002 The first woman elected as President. Prof. Anne Grete Hestnes takes over office on January 1st, 2002.

ISES participates in the World Summit on Sustainable Development (WSSD) in Johannesburg, South Africa.

005 Established on December 24, 1954 the society celebrates its Golden Jubilee at the Solar World Congress in Orlando, Florida, USA in August 2005.